How Scientific Instruments Speak

Postphenomenology and the Philosophy of Technology

Series Editors: Robert Rosenberger, Peter-Paul Verbeek, Don Ihde

As technologies continue to advance, they correspondingly continue to make fundamental changes to our lives. Technological changes have effects on everything from our understandings of ethics, politics, and communication, to gender, science, and selfhood. Philosophical reflection on technology can help draw out and analyze the nature of these changes, and help us to understand both the broad patterns of technological effects and the concrete details. The purpose of this series is to provide a publication outlet for the field of philosophy of technology in general, and the school of thought called "postphenomenology" in particular. The field of philosophy of technology applies insights from the history of philosophy to current issues in technology and reflects on how technological developments change our understanding of philosophical issues. Postphenomenology is the name of an emerging research perspective used by a growing international and interdisciplinary group of scholars. This perspective utilizes insights from the philosophical tradition of phenomenology to analyze human relationships with technologies, and it also integrates philosophical commitments of the American pragmatist tradition of thought.

Recent Titles in This Series

Technology and Anarchy: A Reading of Our Era, by Simona Chiodo

Feedback Loops: Pragmatism about Science & Technology, edited by Andrew Wells Garnar and Ashley Shew

Sustainability in the Anthropocene: Philosophical Essays on Renewable Technologies, edited by Róisín Lally

Unframing Martin Heidegger's Understanding of Technology: On the Essential Connection between Technology, Art, and History, by Søren Riis, translated by Rebecca Walsh

Postphenomenological Methodologies: New Ways in Mediating Techno-Human Relationships, edited by Jesper Aagaard, Jan Kyrre Berg Friis, Jessica Sorenson, Oliver Tafdrup, and Cathrine Hasse

Animal Constructions and Technological Knowledge, by Ashley Shew

Using Knowledge: On the Rationality of Science, Technology, and Medicine, by Ingemar Nordin

Postphenomenology and Media: Essays on Human–Media–World Relations, edited by Yoni Van Den Eede, Stacey O'Neal Irwin, and Galit Wellner

How Scientific Instruments Speak: Postphenomenology and Technological Mediations in Neuroscientific Practice, by Bas de Boer

How Scientific Instruments Speak

Postphenomenology and Technological Mediations in Neuroscientific Practice

Bas de Boer

LEXINGTON BOOKS
Lanham • Boulder • New York • London

Published by Lexington Books
An imprint of The Rowman & Littlefield Publishing Group, Inc.
4501 Forbes Boulevard, Suite 200, Lanham, Maryland 20706
www.rowman.com

6 Tinworth Street, London SE11 5AL, United Kingdom

British Library Cataloguing in Publication Information Available

Library of Congress Cataloging-in-Publication Data

Library of Congress Control Number: 2020949871

ISBN: 978-1-7936-2784-1 (cloth)
ISBN: 978-1-7936-2785-8 (electronic)

Contents

Acknowledgments

Some of the chapters in this book are revised versions of articles that were published earlier. Permission was obtained to use (parts of) these articles in this book.

Chapter 1 is a revised version of a paper coauthored with Hedwig te Molder and Peter-Paul Verbeek, "The Perspective of the Instruments: Mediating Collectivity," *Foundations of Science* 23 (2018): 739–55.

Parts of chapter 4 are derived from a paper entitled "Gaston Bachelard's Philosophy of Science: Between Project and Practice," *Parrhesia* 31 (2019): 154–73.

Parts of chapter 5 are derived from a paper coauthored with Hedwig te Molder and Peter-Paul Verbeek entitled "Understanding Science-in-the-Making by Letting Scientific Instruments Speak: From Semiotics to Postphenomenology," which is forthcoming in Social Studies of Science.

Chapter 7 is a revised version of a paper coauthored with Hedwig te Molder and Peter-Paul Verbeek, "Constituting 'Visual Attention': On the Mediating Role of Brain Stimulation and Brain Imaging Technologies in Neuroscientific Practice," *Science as Culture*, doi: 10.1080/09505431.2019.1710739, © Taylor & Francis, 2020.

Chapter 8 is a revised version of a paper coauthored with Hedwig te Molder and Peter-Paul Verbeek, "'Braining' Psychiatry: An Investigation Into How Complexity Is Managed in the Practice of Neuropsychiatric Research," which is currently under review.

I would like to express my thanks to my colleagues of the Department of Philosophy at the University of Twente for forming my thinking. Especially I would like to thank Peter-Paul Verbeek, for his enthusiasm and support during the writing of this book, and Hedwig te Molder, who introduced me to the fascinating world of ethnomethodology and conversation analysis.

Other special thanks go to Jonne Hoek and Olya Kudina: our continuous discussions and collaborations are an important source of inspiration for my work. Furthermore, I have greatly benefited from discussing my work with colleagues interested in postphenomenology at various conferences, thanks for that.

The research in this book would have not been possible without the willingness of two laboratories to function as an object of study. I want to thank everyone involved in these laboratories, for allowing an individual weirdly calling himself a "philosopher-anthropologist" to peak into the daily activity of doing scientific research.

Finally, I would like to thank my partner Marleen for her support and love while I was finishing this book and my parents for stimulating my intellectual curiosity from an early age onward.

Introduction

Technological Mediations and (Neuro-)Scientific Practice

In 1665, Robert Hooke's *Micrographia* was published. In the preface, Hooke writes, "By the means of Telescopes, there is nothing so far distant but may be represented to our view; and by the help of Microscopes, there is nothing so small as to escape our inquiry; hence there is a new visible World discovered to the understanding" (1665). Famously, by using a microscope, Hooke could observe the microstructure of organisms, enabling him to claim that all living organisms are essentially composed of cells (Gest 2004).

In 1990, Seiji Ogawa and colleagues published an article in which they proposed functional Magnetic Resonance Imaging (fMRI) as a new technique to image brain activity. They wrote that "the results suggest that *BOLD* contrast can be used to provide *in vivo* real-time maps of blood oxygenation in the brain under normal physiological conditions [. . .] and complements other techniques that are attempting [to measure] regional neural activity" (1990, 9868) (Emphasis in original). This experimental result marked a landmark in the development of the cognitive neurosciences because it provided a scientific instrument that allowed the visualization of brain activity in vivo.

In both of these cases, the development of new observational techniques allowed a range of new phenomena to become candidates for scientific investigation, leading to new ways to understand the functioning of organisms. Through such developments, new scientific objects come into being (e.g., cells), or existing scientific objects transform, when considered in a new light (e.g., cognitive functions such as perception, attention, etc.). This demonstrates how the sciences coevolve with technological developments. But how does a new world—revealed through scientific instruments—enter the domain of human understanding, and what kind of world is this?

To answer this question, one must focus on both the role of scientific instruments in scientific practice and how these instruments shape how scientists observe and interpret the world. In this book, I develop an understanding of scientific instruments not as neutral extensions of the sensory capabilities of scientists but instead as *mediating* the reality that scientists study (Ihde 1990; Verbeek 2005). As will become clear, this implies that investigating the role of instruments[1] in scientific practice entails simultaneously investigating how the reality that scientists investigate is continuously transformed by them.

The focus of this book is on how scientific instruments mediate the reality studied in scientific *practice*. This focus leans on two assumptions: (i) that philosophy of science should focus on scientific practice, not just on scientific theorizing and (ii) that scientific instruments actively shape these practices. In this introduction, I legitimize these assumptions, and throughout this book, I flesh out the consequences of these assumptions for our understanding of science.

THE CONTEXT OF DISCOVERY AND THE CONTEXT OF JUSTIFICATION IN PHILOSOPHY OF SCIENCE

Most twentieth-century (analytic) philosophy of science assumed scientific theories as a central unit of analysis. This focus reflects how Karl Popper treated the distinction between the context of discovery and the context of justification. Popper argues that whereas the former is concerned with the circumstances in which scientific theories develop, the latter deals with their logical examination and justification. According to Popper, only the context of justification should be of interest to philosophers of science. The context of discovery, on the contrary, should be considered as a domain of empirical psychology because scientific discoveries do not follow logical trajectories and are therefore inaccessible to philosophical inquiry. In reference to Einstein, he states that "there is no logical path [. . .] leading to universal laws. They can only be reached by intuition, based upon something like an intellectual love for the objects of experience" (Popper 1959, 31). Based on this distinction, a tradition in the philosophy of science arose that was primarily concerned with scientific theories that could be analyzed in terms of their rational justifiability.

Joseph Rouse (1996) has called attempts for the rational justification of scientific theories the "legitimation project" within philosophy of science that searches to eliminate subjective elements from scientific research. However, as Rouse argues, this perspective neglects that science and the rationality associated with it only exist to the extent that they are connected to human subjects. Therefore, scientific theories and the terms functioning within them cannot be dissociated from human activity. Because of this, a philosophical perspective on science should be able to account for the fact that the objects that the sciences postulate and the terms used in scientific theories refer back to the practical circumstances in which scientific objects are created. Such a perspective challenges the strict distinction between the context of justification and the context of discovery and suggests that a philosophy of science *should* be interested in the scientific practices in which the objects of science come into being. In line with this idea, the philosophy of science developed in this book should essentially be considered a philosophy of scientific *practice* (Ankeny et al. 2011).

INSTRUMENTS IN THE PHILOSOPHY OF SCIENCE: THREE PERSPECTIVES

In recent decades, philosophers of science—explicitly or implicitly influenced by studies of science in practice—have begun to note the role of instruments in science. The works of Bas Van Fraassen, Ian Hacking, and Ronald Giere comprise three of the most well-known philosophical accounts of scientific instruments.

Van Fraassen opens his analysis of scientific instruments with the question of whether they should be understood as *engines of creation* or as *windows into the invisible world.* In the latter case, scientific instruments are passive extensions of the human senses, while in the former case, they actively create new phenomena. Accordingly, the question that Van Fraassen considers to be at the heart of an analysis of scientific instruments is whether instruments provide scientists with *new* phenomena or grant access to phenomena that already existed yet were previously invisible to us.

How we should conceive of the role of instruments in scientific practice depends on what we consider a scientific observation to be. According to Van Fraassen, "observation is perception, and perception is something [only] possible for us, if at all, without instruments" (2008, 93). In his view, "naked" perception is something that is contingent on the physiological make-up of human observers, and even then, it is questionable whether "naked" is an adjective applicable at all to the context of perception, as it implies that it is possible for perceptions to be neither theoretically nor historically laden. From this (empiricist) definition of scientific observation, it can be derived that in Van Fraassen's view, scientific instruments cannot be understood as neutral extensions of our perceptual abilities. Even in his early works, Van Fraassen argues that the human organism is the measure of what can be observed: the "able" in observable refers to our limitations as human beings (e.g., 1980, 17). By definition, then, technologies that augment human perceptive capabilities are not windows into the invisible world but instead *engines of creation* that are constitutive of new phenomena.

However, this is not the only possible understanding of the concept of "observation." In his analysis of microscopic observations, Ian Hacking also extensively discusses the relation between instrumental observations and those made with human senses. He holds that microscopes do not passively extend the human senses, as "observation is a skill" (1983, 180). Hence, it is insufficient to explain the role of the microscope in scientific observation in terms of the neutral extension of our sensory field. As such, he proposes speaking of seeing *with* a microscope rather than seeing *through* it. Instead of allowing every observer to passively observe a microscopic body, Hacking argues that the microscope provides images that can only be recognized by skilled microscopists. Hence, according to Hacking, observations involve more than brute perception, as they are structured in relation to the habitual skills that scientists acquire.

Because the phenomena observed by trained microscopists stubbornly refuse to disappear, Hacking holds that we are justified in accepting that microscopic entities exist independently of the microscope. By extending our sensory field, microscopes provide us with already existing yet previously invisible phenomena that can be seen by a skilled observer. In terms of Van Fraassen's distinction, we can say that Hacking develops an understanding of scientific instruments as *windows into the invisible world.* This contrasts sharply with Van Fraassen's own view that by moving beyond our biological limits, "microscopes [. . .] *create* phenomena, to be accounted for by theories" (2008, 101, my emphasis). However, both accounts conceptualize scientific instruments as active participants in the observation process, regardless of whether they are considered engines of creation or windows into the invisible world.

Yet, neither of the mentioned accounts explains *how* the use of scientific instruments allows us to observe previously unknown phenomena. This theme is addressed by Giere, who explains the coming into being of new scientific phenomena in terms of the physical laws that apply to scientific instruments—laws that differ from those that constitute human vision. He holds that, by working according to distinct physical mechanisms, scientific instruments disclose aspects of the world that cannot be detected with the naked human eye. Accordingly, scientific instruments "typically incorporate a built-in perspective on the world" that differs from ours (Giere 2006, 116). His analyses focus on advanced imaging technologies such as magnetic resonance imaging (MRI) and scanning tunneling microscopy. Scientists can observe different phenomena with these technologies, because they provide them with new perspectives different from the human sensory perspective.

In Van Fraassen's terms, Giere understands scientific instruments as *windows into the invisible world*, because "they are perspectival in that they interact with only restricted aspects of the world" (Ibid. 59). They allow the detection of phenomena that remain invisible to the perspective constituted by human vision. In other words, the world is apparently physically structured in such a way that scientific instruments that possess a certain internal physical design allow us to detect certain physical processes in it. However, the perspective a scientific instrument offers does not itself determine what can be seen but only *how* phenomena become visible to us. How detected phenomena are subsequently modeled and interpreted is not guaranteed in the internal design of the technology. When we model these detections and theorize about them, "we simply cannot transcend our human perspective, however much some may aspire to a God's-eye view of the universe" (Ibid. 15). How we interpret observations made by scientific instruments is thus fundamentally dependent on the observing scientist(s).

These three philosophical perspectives share a focus on how scientific instruments can help in establishing a relationship with the objects they investigate. These are important contributions to a philosophy of science limited to the context of justification, because they ask whether it is rationally warranted to believe in the existence of phenomena unobservable by human perception alone. However, when accepting that the division between the context of discovery and the context of justification is not clear-cut, a philosophy of scientific instruments should also focus on how the objects that scientists reason about are created in scientific practices in which scientists relate to reality through technologies.

POSTPHENOMENOLOGY AND THE TECHNOLOGICAL MEDIATION APPROACH

In the philosophy of technology—contrary to the philosophy of science—the question of how human beings relate to the world through technologies has been widely discussed, particularly from the perspective of postphenomenology—or technological mediation theory (e.g., Ihde 1979, 1990; Verbeek 2005, 2011). The point of departure of this line of thinking is phenomenological. Building on the work of Martin Heidegger, technologies are understood as actively shaping how human beings experience reality and the goals they wish to pursue instead of as neutral means that enable the pursuit of preexisting human goals. The American philosopher of technology Don Ihde neatly captures this phenomenon in the term *technological mediation* and develops an understanding of technologies as mediating how human beings relate to the world, thereby shaping their experiences and understanding of the world.

This book can be considered as a systematic attempt to use postphenomenology to investigate the technological mediation of reality that is specific to the practice of science. I intend to show that a postphenomenological understanding of scientific instruments as mediating technologies helps recognizing their crucial importance in scientific practice. Simultaneously, however, I will critically scrutinize the (implicit or explicit) assumption in postphenomenology that technological mediations of instruments in science can be analyzed similar to technological mediations in other practices (e.g., Forss 2012; Ihde 1991), an assumption that downplays the importance of theory in science (e.g., Ihde 1991, 100). I do so by bringing postphenomenology in conversation with several twentieth-century theorists of scientific practice, most notably with Martin Heidegger, Gaston Bachelard, Bruno Latour, and Harold Garfinkel. Accordingly, besides highlighting and analyzing the role of instruments in scientific practice, a key aim of this book is exploring whether a continuity between scientific and everyday practices can indeed be straightforwardly assumed, such that the theoretical nature of science can be analytically surpassed.

SCIENTIFIC INSTRUMENTS IN NEUROSCIENTIFIC PRACTICE

Studying how technologies mediate the reality studied in scientific practices requires considering how scientists interact with technologies. This has been one of the central concerns of the field of science and technology studies (STS), which stresses that relations between humans and technologies should be examined not in isolation but rather against the background of the fundamental collectivity of scientific practice (e.g., Coopmans et al. 2014; Vertesi 2012). When seriously considering this idea, one must ask how the reality that scientists study becomes present in relation to technologies and study empirically how this process occurs within scientific collectives. While studies in STS importantly highlight that the constitution of scientific knowledge must be considered relative to the specificities of scientific collectives, they typically ignore the mediating role of scientific instruments in the constitution of the collectivity of science. In postphenomenology, the collective nature of scientific activity is often neglected, yet cannot be surpassed when investigating how technological mediations are appropriated by groups of scientists (Verbeek 2016). For this reason, an investigation into the mediating role of instruments in scientific practice requires a combination of postphenomenology and STS.

A study of the appropriation of technological mediations remains incomplete when approached philosophically and requires to investigate empirically the role of instruments in scientific practice. The empirical study of scientific (laboratory) practices has been a central interest of STS from its early days onward (e.g., Knorr-Cetina 1999; Knorr-Cetina and Mulkay 1983; Gilbert and Mulkay 1984; Latour and Woolgar 1986). A primary achievement of STS was the stream of evidence it produced against the idea that science has a "view from nowhere" from which a relation between a disembodied epistemic subject and an equally isolated object can be established. Studies in STS instead showed that scientific facts and objects are the products of local, collective efforts and are contingent on active human manipulation.

In the two historical examples at the beginning of this introduction, we saw that. Hooke attributed great importance to the microscope and telescope for the progress of science, and similarly fMRI is attributed great importance in the development of the cognitive neurosciences. However, such historical anecdotes about the importance of scientific instruments do not reveal much about what they do in scientific *practice* and how they shape the reality that scientists investigate. Building on the empirical perspective of STS, I study in this book how scientific objects and facts come into being in relation to scientific instruments. The mere fact *that* scientific instruments are crucial within science does not explain *how* this role must be understood. This requires an empirical investigation of how scientific instruments are adopted within collectives of scientists.

Investigating the role of technologies in scientific practice requires a focus on a specific type of practice. In my case, the focus is on an area of science that has been rapidly growing in recent decades: the (cognitive) neurosciences. I have two primary reasons for focusing on this specific area. Firstly, the development of the neurosciences into a "Big Science" ran parallel with the development of advanced imaging technologies allowing the visualization of brain activity. This makes the neurosciences an excellent area to study how the objects that neuroscientists investigate are shaped in relation to technologies. Secondly, the neurosciences currently draw much public attention and media coverage, often conveying the message that mental phenomena are increasingly demystified by the neurosciences and can be explained purely in materialist terms. The enormous public interest in scientific explanations of the mind makes some suggest that we are currently witnessing a "neurohype" (e.g., Ali et al. 2014). As such, it is pivotal to understand how the objects that neuroscientists investigate and make claims about are shaped by the technologies they use.

STRUCTURE OF THE BOOK

The book is divided into two parts: part I (Toward a Theory of Technological Mediations in Scientific Practice) is a philosophical exploration of how the reality that scientists investigate can be understood as technologically mediated. In it, I show how a postphenomenological perspective on scientific instruments must be augmented to understand how scientific instruments mediate the reality that scientists study. In part II (A Postphenomenological Ethnomethodology of Neuroscientific Practice), I develop an ethnomethodological approach based on the philosophical explorations in part I. This approach is applied to two empirical case studies in which I investigate how knowledge of human cognition in the cognitive neurosciences is mediated by brain imaging and brain stimulation technologies. Part II can be regarded as an empirical translation of the theory of technological mediation in scientific practice developed in part I.

In chapter 1, I show that scientific instruments can best be understood as mediating technologies. I argue that the two of the most promising philosophical attempts to understand the role of instruments in science—Davis Baird's *Thing Knowledge* and Ronald Giere's *Scientific Perspectivism*—mistakenly assume (i) that technologies in science have a determined function and (ii) that the human members of scientific collectives potentially have immediate access to this function. This is illustrated with an example of the use of fMRI in the neurosciences. Using this example, I argue that fMRI constitutes a field of opportunities within which neuroscientific interpretations of cognitive functions can occur. These interpretations are themselves not presupposed in the field of opportunities offered by fMRI. Because of this, neither can the human members of scientific collectives have immediate access—through fMRI—to their object of interest nor do scientific technologies warrant interpretative stability within or across scientific collectives. I suggest that an understanding of scientific instruments as mediating technologies is not committed to Baird's and Giere's assumptions and therefore comprises a promising starting point for understanding the role of instruments and science. Thus far, no systematic account of technological mediations in scientific practices has been developed; doing so is the goal of the next chapters, which constitute the first part of this book.

In chapter 2, I take a detour through the philosophy of technology to explicate how a philosophy of technological mediation leads to a specific perspective on the study of scientific practice. To do so, I discuss how the empirical turn in the philosophy of technology, in general, and the focus on human-technology relations in postphenomenology, more specifically, can be considered attempts to overcome the uniformity of "classical" philosophers of technology. I show that the concept of "technological mediation" is central to the development of a postphenomenological philosophy of technology. By tracing this idea back to Martin Heidegger's famous distinction between the *ready-to-hand* and the *present-at-hand*, and by showing how Heidegger's phenomenology of tool use is transformed into a postphenomenology of human-technology relations, I flesh out the specific meaning of how the term "technological mediation" is used in this book. This is done by showing the relevance of the postphenomenological criticism on the late Heidegger's philosophy of technology that is often dismissed as "too pessimistic" and "too monolithic."

In chapter 3, I show the sense in which the postphenomenological approach demands a modification when moving from the study of human-technology relations in everyday practices to studying them in scientific practices. I argue that Heidegger's hermeneutics of science cannot be dismissed on the same grounds as his analyses of technology. In scientific practices, the explicit aim is to establish a relation with the *present-at-hand* that presupposes a *deliberate rupture* with (the objects of) ordinary experience. This suggests that a central concern for the study of scientific instruments in scientific practice is precisely how a theoretical mode of world disclosure arises from the relation with the *ready-to-hand.* I argue for a revitalization of Heidegger's hermeneutics of science and show that while Don Ihde's philosophy of scientific instruments rightly foregrounds the importance of technologies in scientific practice, it fails to take into account that engaging in scientific practice requires establishing a theoretical relation with the world. Building on Heidegger's hermeneutics of science, I argue that the theoretical character of scientific practices should—despite the importance of technologies—not be sidestepped.

Heidegger suggests that a turnover of our initial relation with the world can occur in relation to technologies but does not offer a clear picture of how this rupture must be understood. In chapter 4, I turn to the French epistemologist Gaston Bachelard to better understand how this rupture is realized. I use Bachelard's term "phenomenotechnique" to clarify how an initial relation with the world is turned over in scientific practice. This allows me to develop an understanding of scientific objects as *artificial* objects that only exist insofar they are realized in the relation between existing concepts, mathematical techniques, available technologies, and experimental setups. When examining scientific practices through this lens, it becomes evident that scientists do not explain or make predictions about an external nature but about artificial objects realized in a phenomenotechnique. I suggest that only when considering this does it become possible to articulate what is specifically "scientific" about scientific practices. This is an important augmentation of postphenomenology: the intentional relation between scientists is mediated less by "material technologies" but rather by a phenomenotechnology (i.e., by the structure within which scientific objects are realized).

Bachelard uncritically assumes that there is a difference between scientific and ordinary practices. In chapter 5, I introduce the work of the French philosopher-anthropologist Bruno Latour to critically scrutinize this strict distinction and highlight that, to explain the coming into being of the reality of science, the mundane practical aspects of science are crucial. According to Latour, any a priori acceptance of a fundamental difference between scientific and other practices is fictitious because the rationality of science is not the prerequisite for, but the outcome of, specific practices. To investigate how the "true" and the "real" come into being, Latour proposes beginning to study science-in-the-making, a term he develops to denote the practice of science before scientific facts have been accepted. Studying science-in-the-making serves to reveal the hybrid nature of scientific facts and objects by laying bare the practices of mediation within which they arise. It is through these mediations that networks are created, within which scientific facts are accepted and acquire the status of "rational" and "true," while they are actually the products of actions of the relevant actants.

Latour crucially highlights the importance of practical aspects in the construction of scientific facts and rightly stresses that the situated practical character of scientific practice demands the empirical study of specific scientific practices. However, Latour's approach can be criticized from the perspective developed in the first part of this work. By contrasting Latour's notion of mediation with the notion of technological mediation as developed in this book, I argue that understanding how technologies allow scientists to relate to new objects requires focusing on how the intentionality of scientists is specifically shaped in relation to technologies. I suggest—contrary to Latour—that this requires understanding the networks of actants as embedded in the project of disclosing reality theoretically *such that* it becomes accessible for scientific inquiry. Accordingly, an empirical approach to studying scientific practices should be developed that considers the theoretical character of the practical, situated activities of scientists.

Part II is concerned with the development and application of an empirical approach to scientific practices that consider the insights obtained in the first part of this book. I develop an approach to the mediating role of instruments in scientific practice that considers both the distinctly *theoretical* (e.g., scientific) character and the non-distinct *practical* character of scientific practice. I propose studying the appropriation of technological mediations in science in terms of the genesis of a collective epistemic stance. Since the specifics of this genesis can only be studied empirically, I use this approach in two empirical case studies focusing on the neurosciences.

In chapter 6, I show that postphenomenological studies of scientific practice to date have placed a strong focus on how technologies mediate how the phenomena that scientists study become perceptually present. The hermeneutic aspects of technological mediations have accordingly been analyzed in terms of the interpretative flexibility instantiated by the gestalt quality of the visual presentation of scientific phenomena. I argue that it is necessary to augment this focus on perception with an analysis of how scientific objects appear in a conceptually meaningful space, such that a shared epistemic stance can be maintained. Phenomenologically inspired approaches in sociology (e.g., ethnomethodology [EM], conversation analysis [CA]) have explored how a shared epistemic stance comes into being in terms of how an intersubjective reality is practically accomplished. Integrating such perspectives into postphenomenology allows the study of how a shared reality comes into being in the practical interactions between scientists and technologies and in the interactions between scientists. This combination of postphenomenology and EM makes it possible to explore how the practical appropriation of technological mediations also shapes the concepts used in scientific theorizing and can help to unravel the conceptual and perceptual strategies that scientists employ to attribute meaning to the phenomena they investigate.

In chapter 7, I use the approach developed in chapter 6 to empirically study how the scientific object "visual attention" in the cognitive neurosciences is shaped through appropriations of technological mediations. With the development of brain imaging technologies in the 1980s and 1990s and the use of noninvasive brain stimulation (NIBS) since the early 2000s, cognitive neuroscientists have developed ways to study the presumed causality of brain-behavior relationships in vivo. Based on a qualitative study of a group of cognitive neuroscientists who combine brain imaging and brain stimulation techniques to establish causal brain-behavior relationships, I show that visual attention is constituted through how scientists manage a tension between the epistemic norms of "causality" and "reality" and that how researchers orient themselves to these norms is mediated by the different technologies used. I argue that NIBS technologies introduce a strong causal imperative that introduces a cause-effect model of the relationship between brain and behavior and that this normative orientation is taken for granted in the scientists' research routine. This leads to a paradoxical situation in which a causal imperative is both a dominant assumption and the desired outcome of research in the cognitive neurosciences. Furthermore, I discuss what implications these results have for how the brain and mind are investigated in big science projects.

In chapter 8, I turn toward another area that uses the methods of neuroscience to study the human mind: neuropsychiatry. Brain imaging technologies in psychiatry are claimed (i) to ultimately offer an objective foundation for diagnostic processes and (ii) to be able to prescribe forms of clinical (pharmacological) treatment that specifically target the symptoms of a specific mental disorder. I show how brain imaging technologies give rise to a conception of the brain as a complex network, which conflicts with the aim to develop straightforward causal explanations of mental disorders. I argue that the constitution of mental disorders in neuropsychiatric practice occurs by managing a potential tension between the epistemic norms of "complexity" and "simplicity." Subsequently, I show that this tension mediates how the adequacy of earlier psychiatric diagnoses is evaluated and how experiments in neuropsychiatry are designed. My analysis suggests that neuropsychiatrists primarily appropriate brain imaging technologies to allow the mapping of cognitive functions to specific brain areas, instead of using them to orient to the brain as a complex network. To conclude, I discuss the implications of this specific appropriation of brain imaging technologies for how mental disorders are conceptualized.

In the conclusion chapter, I discuss how a philosophy of technological mediation must develop in a productive dialectic with the empirical study of scientific practices and how scientific practices can be critically approached from the perspective of a philosophy of technological mediation. On the one hand, I show how empirical case studies challenge the presuppositions of a philosophy of technological mediation; on the other hand, I argue that a philosophy of technological mediation—which explicitly aims to closely engage with scientific and technological practices—*does* allow for developing a critical position with regard to the practices it investigates and I clarify how this can be done with regard to the cognitive neurosciences.

Technoscientific developments are said to develop at such a pace that reflections on them are in danger of running behind. The philosophy of scientific instruments developed in this book attempts to keep pace with technoscience so as to critically engage with it.

NOTE

1. Throughout this book, I use the words "scientific instruments" and "(scientific) technology"—unless specified otherwise—interchangeably. The potential relevance for specifically conceptualizing scientific *instruments* as scientific *technologies* is specified in paragraph 3.4.

Part I

TOWARD A THEORY OF TECHNOLOGICAL MEDIATIONS IN SCIENTIFIC PRACTICE

Chapter 1

Scientific Instruments as Mediating Technologies and the Collectivity of Scientific Practice

Mediating Collectivity

INTRODUCTION: SCIENCE AND SCIENTIFIC COLLECTIVES

It is common knowledge in STS and the philosophy of technology that the collective and technological nature of scientific practice has consequences for how scientific knowledge is produced (e.g., Knorr-Cetina 1999; Latour 1987; Lynch 1994). However, the importance of these findings has only slowly been acknowledged in the philosophy of science. Two attempts that try to integrate these findings into the philosophy of science are Ronald Giere's *Scientific Perspectivism* (2006) and Davis Baird's *Thing Knowledge* (2004). While they diverge in their specific approaches regarding the role of technologies in scientific practice, Giere and Baird share the intuition that the production of scientific knowledge is not something that is exclusively limited to the human realm and that technologies actively shape how scientists understand the world.

Both Giere and Baird adopt as a starting point that the importance of scientific instruments becomes immediately clear when examining scientific practice. According to Giere, scientific knowledge is the product of a network of scientists using instruments that offer a specific perspective on the phenomenon being studied. These instruments are critical in structuring these networks; they offer a solid perspective dependent on their internal design.

The central point of Baird's philosophy is that "instruments, just as theories, bear knowledge" (Baird 2004, xvii). Baird's central claim is that scientific instruments have an important epistemological function in guiding scientific developments. Contrary to Giere's idea, this occurs not because they are used in a specific way, or offer a certain perspective on the world, but because scientific instruments are instances of solidified knowledge. Instruments can be detached from the context in which they develop and can be used by other scientists who can learn new things from them (Ibid. 119). Their solidity allows different scientists to encounter a stable phenomenon, which makes scientific instruments essential in the development of scientific knowledge.

In this chapter, I argue that both Baird and Giere problematically assume (i) that scientific instruments have a universally determined function and (ii) that all human members of a scientific collective[1] have immediate access to this function. However, empirical research on the impact and use of concrete technologies in STS and the philosophy of technology suggests that the function of technologies varies across cases and that the specific nature of this function depends on how it is integrated into the lifeworld of users. Postphenomenology is a movement aiming to develop a philosophy of scientific practice that is most closely tied to this line of empirical research (e.g., Friis 2012; Ihde 1991, 1998; Olesen 2012; Rosenberger 2008, 2011; Rosenberger and Verbeek 2015; Verbeek 2005). A central idea of this movement is that technologies mediate our relations with the world. In this view, scientific instruments are not neutral intermediaries but co-determine how the world is revealed to scientists. However, this active shaping cannot be cut loose from the scientists who use this instrument, which is captured in the postphenomenological notion of technological mediation.

The idea of technological mediation entails that nothing exists prior to the relation between scientist (human) and instrument (technology). In this chapter, I contrast this idea with Baird's and Giere's analyses of scientific practice, in which the internal design of a technology is decisive in how a scientist relates to the world. In doing so, I aim to specifically clarify how a philosophy of scientific practice grounded in human-technology *relations* differs from one in which the importance of either humans or technologies is prioritized. From a postphenomenological perspective, neither scientist nor technology is the decisive factor determining how the world is revealed.

But how can a scientific instrument offer a collective, shared perspective or a clear function when those are not predetermined by the design of the instrument? The postphenomenological focus on human-technology relations has not yet formulated an explicit answer to this problem.[2] In this chapter, I argue that to understand the collectivity of scientific practice from a postphenomenological perspective, it does not suffice to understand the specific individual relations between observers and scientific instruments. The relations *between* scientists should be considered as well. Scientific instruments are crucial in structuring these relations, as all involved scientists necessarily must relate to them. In this relation, it becomes clear what counts as collective knowledge and what is subsequently acted upon (cf. Latour 1987; Hutchins 1995).[3]

The idea that scientific practice must be interpreted in terms of human-technology relations is compared with Latour's actor-network theory (ANT) account of collective scientific knowledge. I argue that in ANT, it is taken for granted that a group of scientists can have a clearly determined perspective with regard to an instrument. From the perspective of technological mediation, however, how it is possible that scientists engage in a similar relation with a scientific instrument is precisely the matter at stake. Furthermore, I argue that the collective relations between scientists and instruments are not presupposed or immediately given but that they are the product of negotiations between scientists.[4] In other words, only in developing a relation with an instrument that is collectively workable does it becomes possible to produce scientific knowledge.

1.1 SCIENTIFIC INSTRUMENTS AS SOLIDIFIED KNOWLEDGE

Baird's aim is to do justice to the importance of instruments in science and to connect this importance to the epistemic role of these instruments. His primary argument is that, just as scientific theories, the instruments used in science bear knowledge that is independent of the presence of an individual observer. Accordingly, he argues that there is something *in* the instrument that is of epistemological significance (Baird 2004, 4). Baird's notion of "instrument" includes a wide array of devices, ranging from telescopes to Boyle's air pump and Watson and Crick's material realization of the Double Helix. However, he does not aim to develop a generic framework capable of explaining the epistemic role of each instrument through a similar mechanism: "Different things, and even different aspects of the same thing, operate epistemologically in different ways" (Baird 2003, 40).

Even though different things have different specific epistemic roles, they have in common that they are all instances of solidified thing knowledge. According to Baird, this type of knowledge is restricted to the domain of things and should be distinguished from our subjective (propositional) knowledge of the world. Contrary to the knowledge of the scientist (the subject), thing knowledge is objective knowledge, as "the epistemological world of science and technology is too big for a single person to comprehend" (Ibid. 44). It is only in things that the knowledge of the scientific community remains solid. Hence, according to Baird, the knowledge of the scientific community is kept together not by the scientists but by the instruments.[5]

Although Baird recognizes that different instruments have different epistemic roles, he does not situate these roles in relation to the users of the instruments (cf. Pitt 2007, 53). In Baird's view, scientific instruments are in fact black boxes, and it is in this aspect that he identifies their epistemic function. For example, it is because a radiologist need not possess a precise understanding of the working of MRI that (s)he is capable of using it. The facts that misdiagnoses still appear and that medical images can give rise to different conclusions is unimportant for Baird in this context, since "there can be no question that our diagnostic capabilities are vastly improved with these new black boxes [new imaging technologies]" (Baird 2003, 59).

In Baird's view, an MRI scan would be classified as a specific kind of thing knowledge—encapsulated thing knowledge (Baird 2004, 68)—indicating that an MRI scan offers a representation of the brain as well as it embodies the physics of magnet resolution. The scanner bears knowledge of magnet resolution physics, and this knowledge can be studied by investigating the specific nature of the working of the scanner. But, in addition, MRI scans are, for example, used to obtain knowledge of the brain; knowledge that is external to the working of the scanner. From Baird's perspective, both are instances of objective knowledge: the MRI bears knowledge of both the human brain and the physics of magnet resolution independent of the presence of an observer.

However, a brain scan can neither be made nor interpreted without the involvement of at least two human individuals: a scientist and a subject lying in the scanner. As a consequence, the epistemic role of an instrument only becomes visible when users relate to it. Consider the following example to illustrate this: assuming that all cars work in a similar way, we would not need (and to some extent we actually *do* not need) precise knowledge of the exact working of a car. However, the practical situation is profoundly different; traffic rules are not part of the internal structure of the car. In the relation between the driver and the car, the decision of how to interpret the traffic rules is made. Similarly, it is in the relation between a neurologist and MRI that an interpretation of a brain scan is made and in which the epistemic role of an instrument is revealed.

Baird is not entirely unsympathetic to the idea that something relevant occurs when an instrument is used by a scientist. He suggests that things can be "read" and that they "enlarge our ability to bring our cognitive apparatus to bear on the world" (Baird 2004, 40). However, how things are put to use remains unaddressed in his analysis. In the end, a gap remains between the objective knowledge present in the things and how scientists obtain knowledge through the use of scientific instruments. This gap is simply not there in scientific practice; in practice, scientists and instruments do interact to obtain scientific knowledge. As we demonstrate, the outcome of this interaction cannot be exhaustively determined with reference to what is "in" the scientific instrument; it is the product of the relation between the scientist and instrument. This change of focus enables us to ask how it is that scientists and instruments can interact such that scientific knowledge can originate within a particular practice.

1.2 SCIENTIFIC INSTRUMENTS AS OFFERING PERSPECTIVES ON REALITY

The interaction between scientist(s) and instrument(s) is a focal point of interest for Giere. Giere's focus on the role of instruments in scientific practice functions to be able to subscribe to the *contingency thesis*, which states that "reality seems capable of sustaining more than one account of it" (Giere 2006, 8) without embracing social constructivism. His starting point is that different scientific instruments—when put to use—offer different perspectives on reality. When using scientific instruments to detect phenomena that cannot be observed with the naked eye, these can only be examined from the perspective of the instrument used. And, precisely because we cannot cut ourselves loose from this perspective, it makes no sense to think of science as an attempt to grasp the objective structure of an external reality. The strongest claim a scientist can make is that "according to this highly confirmed theory (or reliable instrument), the world seems to be roughly such and such" (Ibid. 6).

In Giere's account, technologies are decisive in adopting this or that perspective: "these artifacts [scientific technologies] typically incorporate a built-in perspective on the world" (Ibid. 116). Which specific perspective a technology offers is, in Giere's view, determined by its internal design. However, this does not explain why scientific instruments offer perspectives instead of establishing a one-on-one relationship with the world out there. According to Giere, the perspectivism comes in because scientific instruments—just as human vision—obey physical laws. Telescopes and microscopes respond only to electromagnetic radiation, just as the human visual system does. As a consequence, all of these systems are blind to cosmic rays and neutrinos and are incapable of seeing the particles that constitute, for example, a tree (Ibid. 42). Apparently, the world is physically structured in such a way that specific technologies, which have a specific internal design, allow us to detect certain physical processes and blind us to others.

How the world becomes visible through the interaction between scientist and instrument, in Giere's view, is hence dependent on the interaction between the physical constitution of the world, the physical constitution of the instrument, and the physical constitution of the human visual system. For example, MRI scans of the brain are the result of atomic interactions on a quantum mechanical level, and despite the fact that the structure of the image crucially depends on the choices made on which parameters to measure and how the data is analyzed, our access to the brain is the function of how the MRI scanner generates a certain output (a brain scan) on the basis of the input it gets (Ibid. 56).

Brain scans are therefore not photographs of the brain but offer a view onto the brain from the perspective of an MRI scan (Roskies 2007). For example, electroencephalography (EEG) offers a different perspective on the brain. It is in this sense that Giere understands the contingency thesis: the account of reality that science provides depends on the perspective offered by the scientific instrument. Still, scientists know that they can draw certain conclusions from MRI scans that they cannot from EEG graphs, and vice versa. In choosing between these technologies, scientists decide which perspective to adopt in relation to the goals they have and to the aspects of the brain they wish to highlight. Despite it being impossible to exceed the perspective of one specific type of technology, it is possible to compare the results of these technologies and decide which is most likely to suit one's goals. Because of their relation to specific goals, perspectives are not rigid, and they change over time, either with the introduction of new technologies or by the combination of existing perspectives. However, as Giere stresses, the combining of different perspectives can never eliminate perspectivism; multi-perspectivism does not transcend perspectivism.

In this view, scientists have an active role in determining which perspective to adopt at which moment. Giere summarizes this in the formula "S uses X to represent W for purposes P. Here S can be an individual scientist, a scientific group, or a larger scientific community. W is an aspect of the real world" (Giere 2006, 60). But where do the specific purposes of the scientist come from? Giere attempts to explain these purposes by expanding the notion of perspective from scientific technologies to the level of scientific theories. He argues that theories are sets of models of the world and that the matching of the models with the world can be tested through the provision of scientific instruments (Ibid. 61). Hence, a continuity exists between the perspective offered by an instrument and the reliability of the consequences of having a specific theoretical perspective, which can only be addressed through the use of a scientific instrument.

Scientific knowledge of the world is therefore always the function of the relation between the theoretical perspective of a scientist or a scientific community and the perspective of a scientific instrument, in which the latter allows for "an intersubjective objectivity in that there is roughly a way something looks from a particular location for most normal viewers" (Ibid. 13). Thus, according to Giere, shared theoretical perspectives can also never escape the relation between the internal design of an instrument and the physical structure of the world to which the instrument obeys. However, as we will become clear throughout this book, if the relation between a scientist and a technology cannot be reduced to the physical structures to which both obey, it is also implied that there is no inherent objectivity in the workings of a technology that is explanatory of intersubjectivity. As I attempt to show in this book, a philosophy of technological mediation can function as a starting point for investigating how a shared epistemic stance with regard to scientific instruments can be developed.

1.3 SCIENCE AS DISTRIBUTED KNOWLEDGE

According to Giere, obtaining scientific knowledge is an activity of a system containing at minimum a human and a nonhuman. Considering that no one is capable of conducting current scientific research on his own and that science always involves relations between humans and scientific instruments, Giere proposes interpreting scientific practice in terms of a distributed cognitive system (cf. Hutchins 1995). Rather than individual scientists, these systems as a whole are responsible for the output of scientific practices. From this perspective, it is no longer necessary to ascribe some form of hyperrationality or other special intellectual capacities to individual scientists. In Bruno Latour's words, "No 'new man' suddenly emerged sometime in the sixteenth century. [. . .] The idea that a more rational mind [. . .] emerged from darkness and chaos is too complicated a hypothesis" (Latour 1986, 1).

Though a critical discussion of the work of Bruno Latour, Giere and Moffatt (2003, 308) argue that distributed cognition is the only plausible explanation that ordinary humans with normal cognitive capacities are capable of doing science. Accordingly, the collectivity of modern scientific systems consisting of both scientists and technologies is of primary explanatory value. In this collectivity, the roles of both concrete technologies and individual human cognition dissolve into the cognitive activity of a system that cannot be explained in terms of its parts. Thus, while Baird explains this success in terms of the objectivity of thing knowledge that exists outside the human subjective domain, Giere attempts to do justice to what counts as scientific knowledge is, in most cases, the outcome of relations between scientists and between scientists and instruments.

CERN's Large Hadron Collider can be used to exemplify this view. Located 175 meters under the ground, this 27-kilometer-long collider was developed to test the predictions of fundamental physical theories and is used primarily for the search of the Higgs boson. This project involves hundreds of scientists and engineers working from different continents, indicating that the experiments performed at CERN can hardly be considered the projects of individual scientists. Rather, it is the complex relationship between different scientists and the instrumentation of the CERN that produces new experimental results. Put in Giere's terms, we should treat the Large Hadron Collider as a distributed cognitive system and attribute cognitive capacity to this system as a whole (i.e., to the total organization of individuals and machinery) (Giere 2002, 5).

But how is knowledge generated by and distributed through these kinds of systems? And how do the human parts of the system relate to the system's nonhuman parts? At minimum, a distributed cognitive system must consist of a human and an external representation of some sort, for example, a symbolic manipulation of an arithmetic operation (say, 4,876 times 8,765), which for most of us is too difficult to perform "in our heads." While such a system is relatively simple, much more complex systems are responsible for obtaining scientific knowledge. Not only are there way more elements involved in the system, but the importance of scientific instrumentation adds to the complexity as well. As demonstrated above, Giere argues that these instruments force us to investigate things from a specific perspective that is determined by the instrument.

Key in Giere's analysis is the idea that scientific knowledge should be interpreted as the *cognitive* output of a system. In the end, these systems "make possible the acquisition of knowledge that no single person, or a group of people without instruments, could possibly acquire" (Giere and Moffatt 2003, 305). Yet, *how* this acquisition can occur remains an open question in his analysis.

1.4 THE POSTPHENOMENOLOGICAL PERSPECTIVE OF TECHNOLOGICAL MEDIATION

Similar to Baird, Giere explains the role of scientific instruments as constituting this or that perspective in terms of the internal design of the instrument. As demonstrated above, in Giere's view, the physical structure of the world is organized in such a way that technologies having a specific internal physical design allow us to detect certain physical processes that scientists can then observe. Such an analysis of the collectivity of scientific practice in fact crucially depends on two assumptions: (i) scientific instruments offer only one perspective on a phenomenon and (ii) all human members of a scientific collective immediately share this perspective.[6]

This section demonstrates, however, that both assumptions are problematic. Giere explains the relation between scientist and technology as a relation between a scientist and the predetermined perspective of a certain technology. But how do scientists come to know what this perspective is, if it is shaped as the function of the relation between the scientist and the technology and cannot exist otherwise? Applied to scientific research collectives, answering this question would imply that we should gain an understanding of how what counts as knowledge is constituted in the relation between scientists and their instrumentation. For example, in the relationship between a scientist and a brain scan, some epistemic decisions are, to some extent, made on the basis of the scan itself. In other words, when using fMRI to access the brain, a researcher cannot but interpret human cognition in terms of brain activity that is represented in terms of the blood flow in a specific brain area. The underlying assumption of this kind of brain research is that blood flow and neuronal activity are coupled in one way or another.

In postphenomenology, this relation between human being and technology is conceptualized as the technological mediation of scientific knowledge (e.g., Ihde 1991, 1998; Verbeek 2005). For example, what is considered knowledge about the workings of the brain is not solely determined by the scientific observer. And neither is this knowledge the consequence of the internal structure of the instrumentation. It is only in the relation between scientists and their instrumentation that knowledge is constituted, such that part of what *counts as* knowledge is to some extent determined by the scan itself. fMRI limits the range of explanations of human behavior (e.g., they do not allow for understanding human behavior in terms of balancing the four human humors), but they require interpretation to be understood.

How the brain is related to through fMRI is not merely an interaction between two physical systems, such that the internal design of the scanner offers a specific perspective on the physical structure of the world, which is accordingly brought within the limits of the human visual system. Seeing brain activity on a brain scan requires the prior active involvement of the researcher, who is actively manipulating the content of the image with an intention to filter and simplify the things that an MRI scanner detects, an activity that is not part of the causal interaction between two physical systems.

The MRI scanner generates an image that makes it possible for a scientist to draw inferences about the working of the brain (cf. Suárez 2004). However, not only the process of construction but also the constructed image itself allows for a variety of interpretations. In Ihde's words, these images are *multistable*; they can be coherently interpreted in several ways. To interpret a brain scan is to determine the relevant features of it: to consider certain aspects as relevant and others as not. A scientist must employ a specific hermeneutic strategy to be able to do so (Ihde 2009, 53; Rosenberger 2008, 72). The relation between the brain scan and the observer is what opens the possibility for deciding what counts as relevant brain activity. The knowledge of the brain is not just determined by the perspective of an MRI scanner that obeys the laws of physics. How a scientist observes and the kind of knowledge of the brain he can obtain are the products of this process of technological mediation.

When a scientist accesses the brain through fMRI, his interpretation of the brain scan cannot be reduced to a scan portraying brain activity in a certain way. When interpreting such a scan, the blue and red dots are related to a certain experimental setup. How the distribution of these dots is interpreted is in itself not determined by the physical design of the imaging technology. The association of brain activity with certain cognitive tasks is not internal to the information present in the brain scan. While this imaging technology makes it possible to interpret human cognition in terms of blood flow in a certain brain area, the meaning of this increase cannot be reduced to how it is measured. For example, when linking activity in certain brain areas to how human beings process visual information, neuroscientists must link the concept of visual processing to the red and blue dots present in the brain scan. Furthermore, the basic assumption underlying the use of fMRI in research in the cognitive neurosciences (i.e., the coupling of cerebral blood flow to neuronal activity) is itself not entailed in the working of the MRI scanner, and neither can this hypothesis be specifically put to test by the scanner. In relating to the MRI scanner, scientists must actively construct a perspective on the brain that necessarily requires the modeling of the relation between blood flow and neuronal activity, which greatly influences how the brain becomes present.

Interpreted in this way, a brain scan does not offer a perspective on the predetermined structure of the brain but makes the brain present in a very specific manner. This translational process is the product of the individual relation between the scientist and the brain scan. Varying on Giere, this can be formalized as "in the relation between S and M, an aspect of W comes into being for purposes P at time T." S refers to an individual scientist having a relation with the mediating technology M, such that an aspect of W (the brain) becomes present to him to investigate this or that phenomenon at a certain point T in time, in which a specific set of scientific technologies is available.

In this view, the brain scan does not stand in between the scientist (subject) and the brain he is studying (object) such that it offers a way to see aspects of the brain that the scientist was previously incapable of. Rather, it is in the relation between human and technology that a scientist can experience the brain in how it is portrayed on a brain scan (Verbeek 2005, 130). In contrast with what Baird and Giere argue, the MRI scanner is not some technology in which a specific perspective or specific kind of knowledge is hidden that can be discovered by a scientist. From a postphenomenological perspective, the working of an MRI scanner cannot be reduced to the physical law that it obeys. When confronted with a brain scan consisting of several blue and red dots, the scientist is not presented with a photograph of the brain, but (s)he must interpret what these dots mean (Carusi and Sissel Hoel 2014). And it is precisely the specific interpretation of this brain scan that is not part of the internal design of the scientific instrument.

1.5 IMMUTABLE MOBILES AND SCIENTIFIC NETWORKS: THE ROLE OF SCIENTIFIC INSTRUMENTS FROM AN ACTOR-NETWORK PERSPECTIVE

Studying how scientists and scientific instruments interact to construct "facts" about "nature" is a primary ambition of ANT, which aims to investigate science-in-the-making. Phenomena and their interpretations are not taken as "given" or "discovered" but as constructions that can be studied while being assembled. In his article "Visualization and Cognition: Thinking with Eyes and Hands" (1986), for instance, Latour develops an account of how knowledge is stabilized in networks of scientists, instruments, and other artifacts. His analysis focuses on how drawings and images are capable of mustering the largest number of well-aligned and faithful allies. One of his most important arguments is that it is precisely these visualizations that make it possible to muster allies, as they allow for showing the whole of the conducted research in one glance (Latour 1986, 5).

Consequently, for Latour, it is not only the perceptual content of the image that is at stake but rather how the content of the image becomes factual because of its ability to muster as many powerful allies as possible. By introducing the concept of *immutable mobile*, he attempts to account for different people drawing similar conclusions when confronted with a specific kind of representation of an object. He states,

> If you wish to go out of *your* way and come back heavily equipped so as to force others to go out of *their* ways, the main problem to solve is that of *mobilization* (Emphasis in original). You have to go and come back *with* the "things" if your moves are not to be wasted. But the "things" have to be able to withstand the return trip without withering away. Further requirements: the "things" you gathered and displaced have to be presentable all at once to those you want to convince and who did not go there. (Ibid. 7)

In Latour's view, therefore, scientific images are in fact to be seen as embodiments of power—comparable to how Giere approaches them as embodiments of knowledge. Images have the power to bring facts into existence. Yet, the question remains whether this analysis exhausts the role of instruments in scientific practice, since it neglects how images are interpreted in the first place and how they inform the interpretational frameworks of scientists. Latour is primarily interested in the extent to which a scientist or a group of scientists is capable of mustering *other* allies and passes over the question of how this group of scientists *itself* comes to treat one potential immutable mobile as more effective than the other.[7]

The absence of this concern is clearly visible in Latour and Woolgar's *Laboratory Life* in which they stress that scientific "phenomena *are thoroughly constituted by* the material setting of the laboratory" (Latour and Woolgar 1986, 64) and that "each item of apparatus [is] combined with certain skills to form specific devices" (Ibid. 69), but do not pay attention to *how* these phenomena are constituted or *what* skills are used to work with the available scientific instruments (Emphasis in original). Instead, they focus on how the material environment of the laboratory dissolves when transformed into a scientific paper. However, scientific instruments, and other technologies such as brain scans, must become immutable mobiles *within* a scientific community as well when they are to be used and interpreted in similar ways.

It is exactly at this point that the postphenomenological approach of technological mediation differs from ANT. Because of its phenomenological orientation, the mediation approach studies the role of technologies in scientific practices "from within." Rather than looking "outside-in" to the formation of networks of relations between scientists and technologies, it looks "inside-out" to the formation of perceptions and interpretational frameworks in relation to the technologies that play a role in scientific practices.

The central ambition is not to "deconstruct" the networks behind scientific facts and theories but to develop accounts of the mediating role of technologies in "science-in-the-making." From a postphenomenological point of view, such accounts can only be studied from a first-person perspective (i.e., by analyzing how scientists develop concepts and interpretations in interaction with the scientific instruments and technologies they use). In other words, the concepts that scientists apply when "reading" a brain scan cannot be reduced to the visual content of the brain scan itself. Establishing a relation between the visual content of a brain scan and a cognitive concept, such as aggression, requires interpreting the visual content in terms of something else (e.g., aggression). Only when a neuroscientist can make this translation do brain scans become relevant for research in the cognitive neurosciences. A phenomenological perspective has the potential to take into account such "interpretational-frameworks-in-the-making." The central question then is not if and how scientific instruments "embody" specific interpretational frameworks or scientific theories but rather how they help such frameworks and theories to come into being.[8]

Scientific instruments help to shape relations between scientists and the reality they study; in doing so, they help to organize how reality is "given" to scientists and how scientists are constituted in relation to the phenomena they study. Scientific instruments, therefore, must be seen as hermeneutic devices (Ihde 1998)—they are "epistemology engines" (Ihde and Selinger 2004)—establishing relations of investigation, in which both the investigator and the investigated are constituted in a specific, technologically mediated way. The event of constitution is what is further explored in this book in terms of the technologically mediated reality made present through scientific instruments, with a focus on how these mediations are appropriated and give rise to collective forms of knowing.

1.6 TECHNOLOGICAL MEDIATION AND COLLECTIVE KNOWING

When it is only in relation to scientific instruments that the objects of research can be observed, instruments function as common denominators to which all scientists within a collective must relate to, both individually and collectively. As I demonstrated above, this has two important consequences: firstly, because scientific instruments are not neutral extensions of human sight but constitute a specific type of the observer and the observed, we should analyze scientific observations in terms of technological mediations.[9] For example, how cognitive phenomena such as aggression must be explained changes when accessed through fMRI, as they become necessarily related to neuronal activity. The linking of neuronal activity to aggressive behavior requires a prior understanding of aggression that is materialized in an extensive experimental setup that presupposes this specific understanding of aggression. This specific prior understanding is in turn reflected in how aggression is visualized through the use of fMRI. Hence, I hold, contrary to Giere's analysis, that the perspective an instrument offers is not merely a matter of the interaction between the physical constitution of a technology and the physical structure of the world.

Secondly, the human parts of a scientific collective must develop a shared epistemic stance towards the mediating technology that allows for multiple coherent interpretations. As for example indicated above, there are several coherent ways to interpret the content of a brain scan. Since no specific perspective is built into an MRI scanner, a workable stance toward the epistemic function of the scanner is something that must be established. This forces the postphenomenologist to develop an alternative answer to the question of how a shared epistemic stance can come into being.

Let me attempt to answer this question through an analysis of an everyday example and by identifying the differences with Baird's and Giere's approaches.[10] Imagine that you must walk from one place to another using a map of the area you are in. The map limits several possibilities; a possible route straight through a dark forest is not indicated on the map, since it only covers registered walking paths, but the map will also not stop you from taking an unregistered route. Moreover, it will not exactly tell you where you are; you must decide that in relation to the map. In other words, both you and the map have an active part in determining the path you will walk. In making certain aspects of the landscape relevant, while neglecting others, the map shapes the landscape and your position in it in a specific way. In other words, in your relationship with the map, the map mediates how the landscape that you are in is revealed to you.

Now, imagine that a group must perform a similar exercise. One or more somewhat adventurous group members offer the suggestion to cross the dark forest even though its structure is not displayed on the map, because they are convinced that it will be a faster route. Only after discussion is this possibility ruled out or accepted by the group as a whole; in this discussion, the members of the group start to display a similar epistemic stance with regard to the map (at least for the moment). Furthermore, in this situation, the map will also not "tell" the group where it is. The location of the group on the map will probably be the result of a discussion as well. In both situations, an epistemic stance is not presupposed, but is the consequence of how the group acts in relation to the mediating technology. The point here is that coming into being of a collective epistemic stance can be studied by analyzing the mediations of which it is the product (Heritage 1984, 259).

Baird would understand the epistemic function of the map as bearing knowledge of the relevant surface. In Baird's sense, this would mean that in the map, there is knowledge of the specific landscape. However, in his model, it remains unclear how, and if, the group can access what is "inside" the map; hence, it remains unclear how the group can obtain knowledge of the terrain through the map. What is at stake is the relation between the instrument and its knowledge of the landscape, without a user entering this epistemological picture. When considering larger collectives such as the one described in the example, Giere suggests that the perspective that an instrument offers is immediately grasped by every member of the collective. Thus, given that the visual capacities of the group members are roughly similar and share similar purposes, they will all immediately grasp how the map relates to the environment where they are.[11] This immediacy is guaranteed by the "intersubjective objectivity" that is presupposed in the perspective offered by the map (cf. Giere 2006, 13). However, he acknowledges that the situation is slightly more complex than noted above when discussing the role of MRI in the neurosciences: "MRI, in particular, makes it abundantly clear that not only is scientific observation perspectival, but also that there are multiple perspectives from which one must choose and no 'objectively' correct choice [can be made]. A lot depends on the goals of the investigation at hand" (Ibid. 56). Considering that scientists determine the goals of their investigations in relation to both each other and the scientific instrumentation, this would already imply a shared epistemic stance with regard to the instrument.

In terms of the map example, Giere understands "the goals of the investigation at hand" as already present before the map is put to use. However, as we have attempted to clarify, it is precisely these things that are still unclear when a map is used. Similarly, how and what can be seen while using an MRI scanner and which experimental goals are developed in relation to MRI are not part of the internal design of the scanner. Of course, our eyes cannot look beyond their visual capabilities, and the working of an MRI scanner must work in accordance with the technicalities of its design. However, collective knowing can never be reduced to these technicalities. As Van Baalen et al. (2017) have shown in the context of multidisciplinary clinical decision-making, collective knowing is always mediated by both the social relations between scientists and the technologies that shape the knowledge that is distributed within a group of scientists. As my example suggests, this applies not only to multidisciplinary clinical contexts but to nonclinical and non-multidisciplinary scientific practices as well.

While scientists use instruments that are capable of detecting and visualizing phenomena invisible to the naked human eye, the instruments do not provide us with an immediate understanding of those phenomena. Giere correctly emphasizes the importance of the former aspect, but he neglects that visualizations do not have a univocal meaning and can be interpreted in multiple ways. And if the goals of the investigation at hand coevolve with the interpretation of the visualization, there is no reason to presuppose agreement with regard to for which goals the visualization should be put to use. Hence, the perspective of the instrument cannot function as offering an objective ground that can establish a collective epistemic stance. Similarly, there is nothing *in* the map that allows the members of the group of travelers to establish an understanding of its precise meaning. As indicated in the above example, the displayed intersubjectivity is the consequence of how the members of this group relate to the mediating technology (in this case, the map) and how this relation structures how the members of the group relate to each other.

1.7 CONCLUSION: MEDIATING TECHNOLOGIES AND SCIENTIFIC COLLECTIVES

The philosopher and historian of science Hasok Chang proposed something along these lines by suggesting that science should be understood in terms of epistemic activities that have a particular aim. According to Chang, it is possible to understand scientific practice in terms of the aims of scientists and the expectations that they possess. However, neither is a standard model of the actors at work within the collective nor are the aims of a collective set from the start, as a scientific collective "only exists because someone upholds it by means of some mechanisms for propagation and maintenance" (Chang 2014, 73). Here, Chang opens the door for understanding the shared epistemic stance of a group of scientists as something that is actively put together by the involved agents rather than presupposing that they share some form of "intersubjective objectivity." This perspective makes it possible to understand how specific aims of a collective are related to the available instrumentation. After all, the agents within a collective are not rigid entities but come into being in relation to the technologies at hand.

Still, the question of how scientists can have a workable shared epistemic stance with regard to an instrument that can potentially generate multiple perspectives remains unclear. It is clear that researchers working at CERN are capable of speaking the same language with regard to a similar set of objects and that neuroscientists can draw meaningful conclusions from brain scans collectively. That the technological mediation of knowledge originates in the individual relation between scientist and instrumentation does not prevent scientific theories from being developed or clinical decisions from being made.

But, when it is not the objective features of the scientific instrument that make it possible to adopt a similar perspective, how then is a shared epistemic stance with regard to the instrument established? Does it presuppose some form of intersubjectivity that makes it possible to use it collectively? These questions can be answered both in the positive and in the negative: in the positive, because an intersubjective epistemic stance seems to exist in scientific practice; after all, how would it otherwise be possible to talk about experimental results? It should be answered in the negative as well; the intersubjectivity is not presupposed. This is not to say that establishing intersubjectivity would be the final goal of such a collective; displaying a shared epistemic stance is the dynamic and preliminary product of the individual relations between scientists and mediating technologies.

In the end, the ongoing development of a workable epistemic stance of a group of scientists with regard to an instrument mediates how the world becomes present to this group of scientists; however, we should interpret these collectives neither as having access to "nature" or to "the world" nor as unchanging systems determining how the world comes into being. An epistemic stance is a fluid state changing over time and relative to the interactions of the members of the collective, rather than a universal decision to use a scientific instrument in this or that way such that the world can be accessed in this particular manner (Te Molder 2015). The shared epistemic stance is not an extension of Giere's idea from the level of scientific instruments that "typically incorporate a built-in perspective on the world" to collectives having a similar built-in design. The crucial difference is that nothing can be built in into such a collective, since the relation between scientists and mediating technologies is not fixed.

This leaves open the issue of how collective scientific practices are structured around individual human-technology relations. As I have attempted to clarify, this collectivity is not something that is an inherent property of the system as such. In stressing that agents are responsible for the propagation and maintenance of a shared epistemic stance, the implication is that the development of an epistemic stance requires action. Such actions cannot be performed outside the technological realm when the mediation role of scientific technologies is considered. The specific nature of those actions is not pre-given but is structured by the relation with a mediating technology. Accordingly, what this mechanism looks like is dependent on the specific relations between scientists and mediating technologies, and the actions that are performed such that a shared epistemic stance is displayed within a scientific collective. In other words, technological mediations cannot be disconnected from how they are appropriated within scientific collectives. A philosophical account of the role of scientific instruments in scientific practice should therefore not be limited to investigating individual human-technology relations but should also be able to explain how a shared epistemic stance comes into being within scientific collectives.

In this chapter, I argued that if instruments in scientific practice are best understood as mediating technologies. While several studies have analyzed scientific practices from this perspective (e.g., Forss 2012; Hasse 2008; Rosenberger 2008, 2013), no systematic account of the role of mediating technologies in science has been developed thus far. In this book, I attempt to develop exactly such an account. In the next chapter, I take a detour through the philosophy of technology underlying the concept of technological mediation to clarify how this account both builds upon and augments a phenomenological perspective on scientific practices.

NOTES

1. In this chapter, I use the notion of "collective" to denote a group of scientists that work in the same laboratory to investigate topic X through the use of specific set of technologies. While I think that a similar objective may apply to Baird's and Giere's accounts if the notion of "collective" would be extended to a larger group of scientists (say, everyone investigating topic X through the use of a specific set of

technologies), I restrict myself to this narrow notion of collective to make as clear as possible that scientific observation are grounded in human-technology relations.

2. This question is related to what Rosenberger (2017) has recently called postphenomenology's problem of invariance. That is, if technologies in scientific practice should be understood in non-essentialist, non-foundationalist terms, how can a (multi-)stable use of them be maintained across individuals?

3. In this chapter, the term "intersubjectivity" is not cognitively interpreted in terms of the mental states of a (set of) individual(s). Rather, it is understood as a shared stance that enables individuals to perform certain actions.

4. Note how this idea differs from the idea that the relations between scientists and instruments are structured through already existing social mechanisms. In my view, this idea mistakenly assumes a social structure that exists outside of human-technology relations. A detailed discussion of this difference is beyond the scope of this chapter. For an analysis of how social relations between scientists change and take a specific shape in different human-technology relations, see Vertesi (2012).

5. In line with Popper's notion of the "third world," Baird argues that scientific instruments bear objective knowledge that may remain undiscovered. For a critical discussion of this aspect of Baird's philosophy, see Kletzl (2014, 199).

6. Giere uses the term "system" to denote a collective of humans and nonhumans, which seems to be closely related to his idea that scientific practices should be understood in terms of distributed cognition. I prefer to use the more general term "collective," because the view of scientific practice developed in this chapter is not necessarily to be understood in terms of distributed cognition.

7. Note that this does not imply that a group of scientists is only interested in mustering allies *outside* science per se, but equally applies to attempts to convince other (groups of) scientists.

8. A more detailed discussion about the similarities and differences between Latour's approach and a postphenomenological one can be found in chapter 5.

9. Vertesi (2012) has recently discussed how human-technology relations shaped the social order in laboratories. The present discussion focuses on how scientific instruments structure epistemic judgments, rather than how they contribute to the social organization of laboratory work.

10. ANT is deliberately omitted from this comparative analysis, as I hope to have shown in the previous section that ANT primarily aims to understand the working of a scientific instrument when a shared epistemic stance is already established.

11. In his discussion of mapmaking, Giere, for example, states that maps are *interest relative*. He argues that road maps are useful to car drivers because they highlight relative distances, but that they are less useful for cyclists because they do not give a clear indication of elevation, which would be of great importance when riding a bike (Giere 2006, 73).

Chapter 2

"Technology" and "Human-Technology Relations"

INTRODUCTION: "CLASSICAL" AND "EMPIRICAL" PHILOSOPHY OF TECHNOLOGY

Before starting to specifically explore the role of technologies in scientific practice, I discuss how their role in daily practice is conceptualized within postphenomenology in this chapter. This functions to (i) provide an introduction to how technologies are understood as mediating reality in the philosophy of technology and (ii) show the phenomenological stance that this understanding of scientific instruments presupposes.

Let me first give an intuitive example of our dependency on technologies. My journey to work is dependent on a large technological infrastructure, which goes largely unnoticed unless specifically reflected upon: my alarm wakes me, and thereafter I use the shower, step on my bike, and cycle to the train station from where I travel to my work by train. Arriving at my work, I use the coffee machine and turn on my computer. E-mails are waiting for response; the information that I need to continue my research I retrieve from papers that are accessible in an online environment. And, occasionally, I write down my thoughts and findings in the word processing program on my computer. In sum, without the presence of an advanced technological system, my daily life could not be structured as it is presently.

Philosophers of technology up until the second part of the twentieth century approached both technology and science as all-encompassing forces that are external to society yet dominantly structure how society is organized (e.g., Ellul 1964; Heidegger 1977a; Jaspers 1951). Martin Heidegger's analysis of technology in "The Question Concerning Technology" is often singled out as the most important example of this "classical" approach in the philosophy of technology and remains an inspirational source for many current philosophers of technology. In this work, Heidegger attempts to find the essence of modern technology and identifies this essence as the disclosing of nature as *standing-reserve*. Technology is interpreted as a mode of being through which the world becomes present as a set of resources at human disposal. According to Heidegger, this prevents human beings from having an authentic relation with reality, and as a consequence with their own being. In this view, the technological infrastructure described above is considered an alienating system that determines how society is organized and how individuals understand themselves.

Such a "classical" philosophy of technology presupposes that different technologies operate according to a uniform scheme and are often referred to as treating technology as a "monolithic phenomenon" (e.g., Feenberg 1999). Technology is considered an alienating force that prevents human beings from having access to their authentic form of being. In these cases, technology is conceived with a capital T and is not sharply distinguished from Science, which was thought of having a similar alienating power. For example, analyses criticized the presence of mass production, in which human beings were interpreted in terms of their production rather than in terms of their human qualities. These kinds of analyses often embody a fear that nature and human beings can only be approached as things that should be dominated or controlled.

At the end of the twentieth century, the philosophy of technology moved away from analyzing Technology as a uniform structural phenomenon that humanity should be guarded against. Instead, philosophers started to investigate how technologies provide human beings with new opportunities to relate to the world rather than how they necessarily alienate human beings from themselves and from nature (Achterhuis 2001; Ihde 1993; Verbeek 2005). This is often referred to as the "empirical turn in philosophy of technology," indicating that the subject matter of a philosophy of technology should no longer be "Technology" but rather "the actual workings of technologies in concrete practices" (Franssen et al. 2016; Kroes and Meijers 2000).

Postphenomenology is an important approach within this empirical turn. It focuses on how reality comes into being in the relations between human beings and technologies. A central idea within postphenomenology is that technologies *mediate* the relationships between human beings and the world, thereby co-constituting specific experiences and understandings of reality. The idea of technological mediation oscillates between technological *determinism* and technological *instrumentalism*: how human beings relate to the world is not determined by the technologies they are surrounded with, but neither are human beings completely free in choosing how the world is revealed to them through technologies.

This chapter specifically focuses on the relation between postphenomenology and the phenomenological hermeneutics of Martin Heidegger to sketch the general philosophical background of the notion of *technological mediation* used throughout this book. On the one hand, postphenomenology dismisses Heidegger's analysis of technology as developed in *The Question Concerning Technology* as treating technology in a monolithic manner. However, on the other hand, postphenomenology is heavily inspired by Heidegger's phenomenology of tool use as developed in *Being and Time*, where he shows that our everyday contact with the world occurs to a large extent in relation to the tools that we employ. Through a discussion of how postphenomenological approaches use Heidegger's early phenomenology of tool use as a corrective to his later analysis of Technology, I clarify the central notion of *technological mediation* that will be central in my analysis of scientific instruments.

2.1 READY-TO-HAND AND PRESENT-AT-HAND: HEIDEGGER'S PHENOMENOLOGY OF TOOL USE

In his phenomenology of tool[1] (*Zeug*) use in *Being and Time*, Heidegger distinguishes between entities that appear as ready-to-hand (*Zuhanden*) and entities that appear as present-at-hand (*Vorhanden*). Heidegger argues that our primary encounter with the world is one in which the things around us are immediately put to use, such that they disappear from view (*ready-to-hand*). He contrasts this with a way of relating to objects as entities with certain qualities that can be described, but are ultimately foreign to us (*present-at-hand*). This distinction appears at the background of Heidegger's criticism of any ontology that takes as a starting point that the world in which we live and act is one where we attain knowledge of the entities around us through giving accurate descriptions of their appearance: "Houses, trees, people, mountains, stars. We can *depict* the way such entities 'look,' and we can give an *account* of occurrences in them and with them (Emphasis in original). This, however, is obviously a pre-phenomenological 'business'" (Heidegger 1985, 91). This is not to say that these descriptions of entities are necessarily incorrect but instead that even if they are, "in conformity with the basic assertions about this entity, which the *mathematical natural sciences* provide, [they] will never reach the phenomenon that is the 'world'" (Ibid., 92, my emphasis). So, if we aim to develop an ontological understanding of the world around us, we cannot turn to the natural sciences, because these remain to consider our relationship with the entities around us in terms of the present-at-hand.

This distinction is influential in postphenomenology (e.g., Ihde 1979, 2010; Verbeek 2005) and is used as an argument that shows that our encounter with the world is always mediated by the technologies we use: the ready-to-hand is attributed ontological priority in how reality becomes present to the experiencing subject. In this chapter, I show how the perspective of technological mediation develops a philosophy of technology based on the ontological priority of the ready-to-hand. However, as becomes clear in the next chapter, placing too much focus on the ready-to-hand neglects that instruments in science also mediate how a relation with the world is revealed in terms of the present-at-hand. Before making this specific point, it is necessary to clarify how—through phenomenological analysis—the ready-to-hand is ascribed ontological priority in Heidegger's thinking.

The Being of Equipment

In his famous analysis of hammering, Heidegger argues that one is no longer aware that one is using a hammer when engaging in the practice of hammering. The hammer becomes ready-to-hand when put it to use, making it disappear in the whole of the context in which one is immersed. When using it, one's concern is not with the hammer, but with something else, such as the nail one is hammering in the wall or with the table that one is making. Accordingly, the hammer becomes transparent; it remains unnoticed in daily practice. The consequence of this phenomenological analysis is that one's first acquaintance with the world is one of immediate relations with tools that are ready-to-hand within a certain totality. This notion of "totality" is phenomenologically important, because one never discovers tools in isolation but is always already immersed in a context that is constitutive of how one relates to tools:

> To the Being of any equipment there always belongs a totality of equipment, in which it can be this equipment that it is. Equipment is essentially "something in-order-to . . ." ["etwas um-zu . . ."]. A totality of equipment is constituted by various ways of the "in-order-to," such as serviceability, conduciveness, usability, manipulability. (Ibid. 97)

Given that the particular way in which one relates to tools always presupposes a context in which they appear as something-in-order-to, the particular way in which a tool can become of service or is put to use cannot be reduced to the particular qualities of the tool that is related to. Both the relationship with a hammer and how it is put to use already presuppose a larger totality of tools in which the hammering occurs. One always finds oneself already in a given context in which the ready-to-handness of tools already is shaped *within and by* this given context. It is through this context and the tools that are encountered within it that the world manifests:

> What we encounter as closest to us (though not as something takes as a theme) is the room; and we encounter it not as something "between four walls" in a geometrical spatial sense, but as equipment for residing. Out of this "arrangement" emerges, and it is in this that any "individual" item of equipment shows itself. *Before* it does so, a totality of equipment has already been discovered (Emphasis in original). (Ibid. 98)

The argument Heidegger makes is thus that one's relationship with the world is not one in which a subject encounters a manifold of objects in geometrical space that appear as in immediate need of specific description. On the contrary, the "arrangement" of the world and how tools reveal themselves in it is constituted by the situation in which one finds oneself hammering a nail in a piece of wood (e.g., because I wish to make a house to find shelter against storm). Or, less primitively, I find myself typing on a computer, which is meaningfully disclosed as something to use as a way to write this book. My primary involvement with this computer is not one of looking explicitly at the specifications of this computer but is made relevant in this larger context.[2] How the world is disclosed to us in this analysis is thus characterized by the *ready-to-hand* manner in which the tools around us appear as relevant as "something-in-order-to" within the totality in which we are immersed.

Breakdown and the Present-at-Hand

The phenomenological observation that the primary encounter with the world is characterized by the ready-to-hand way in which one relates to the things around one does not necessarily rule out the possibility that these things *can* appear as present-at-hand. According to Heidegger, approaching entities in terms of the specific qualities that they have (i.e., in terms of their present-at-handness) does occur in cases of *breakdown*. For example, when the hammer no longer appears as "handy" during the practice of hammering, one starts to conceive of it as an object that has certain qualities (a certain weight, a specific hardness, etc.) (i.e., the hammer becomes present-at-hand as a thing to be studied). Analogously, the specific makeup of the computer that I use only becomes apparent when it no longer functions within this context, and it is no longer possible for me to continue writing this text. As such, the becoming of my computer as an object that is present-at-hand is caused by the breakdown of its initial function within the context I was using it; it was no longer of service within this specific context because its relation with the totality of tools had suddenly disappeared.

In instances of breakdown, "the presence-at-hand of the ready-to-hand makes itself known in a new way as the Being of that which still lies before us and calls for our attending to it" (Ibid. 104). As a consequence, I may ask myself why the computer is no longer working, which may make me interested in the specific weight of the object that can indicate that the motherboard of the computer has dropped out, which may make me conclude that this can be one reason why the thing is no longer functioning as expected, and so on. The computer is only revealed as present-at-hand when it no longer can be used to write this text and as turning over my initial experience of it within a totality of equipment. Entities thus appear as present-at-hand as the consequence of a breakdown of its functioning in a totality that is taken for granted in the everyday contact with the world. Hence, Heidegger is not arguing that the things cannot appear as present-at-hand, but rather that this character is not phenomenologically (i.e., ontologically) constitutive of our relationship with the world.

What is at stake for Heidegger is not primarily giving an "adequate" description of how human beings relate to tools but rather what it means for human beings to be *in the world.* And to understand what this fundamental characteristic of human beings means, it is necessary to understand the phenomenon of worldhood as a state of human beings. When building an ontology grounded in giving accurate descriptions of the objects around us in terms of their present-at-handness, worldhood is surpassed, because

> worldhood itself may have as its modes whatever structural wholes any special "worlds" may have at the time; but it embraces in itself the *a priori* character of worldhood in general. [. . .] The derivative form "wordly" will then apply terminologically to a kind of Being which belongs to Dasein, never to a kind which belongs to entities present-at-hand "in" the world. (Ibid. 93)

This is the crux of Heidegger's argumentation: the world is not something that consists of a set of entities that are present-at-hand outside of human beings, but should be understood as being disclosed *by* them. This is why worldhood is defined as an existential of human beings prior to our encounter with any "world." This world is not something that can be thought of as independent of human beings but is disclosed through the existential structure of human beings, which always implies the a priori character of its worldliness. So, crucially, Heidegger's phenomenology of tool use does not target mathematical and scientific descriptions as being incorrect or not worthy of consideration but functions to note that there is a form of engagement with the world that is prior to its mathematical-scientific disclosure.

2.2 FROM TOOLS TO TECHNOLOGIES: A POSTPHENOMENOLOGY OF HUMAN-TECHNOLOGY RELATIONS

Complex technologies, such as smartphones, and, in the context of scientific practice, fMRI or the Hubble telescope, do not—taken at face value—resemble the tools that Heidegger discusses in *Being and Time*. However, Heidegger's phenomenology of tool use can function as a foundation for analyzing how human beings relate through the world through more advanced technologies. In postphenomenology, this extension is made by replacing the notion of tool with the notion of technology. In this interpretation, a phenomenology of tool use is not limited to primitive tools such as hammers, but allows that more sophisticated technologies can also be analyzed in a phenomenological manner. One can think of fridges, computers, or bicycles, but also of larger technological infrastructure, such as railway systems and the printing press, as ways that shape how human beings relate to the world.

In *Technics and Praxis* (1979), Don Ihde attests to how closely his ideas about the phenomenology of technology are aligned with Heidegger's phenomenology of tool use in *Being and Time*. At the same time, he argues that Heidegger's vocabulary needs some modification when applied to our relations with technologies and that the variety of ways in which the world can be experienced and interpreted through technologies requires a broader framework (Ihde 1979, 4). According to Ihde, Heidegger implicitly assumes that we engage with each tool in a roughly similar manner. One of Ihde's central points is that the specific way in which a technology allows for a relation to the world is not only contingent on the totality in which it appears but also depends on characteristics of the technology itself. Accordingly, Ihde holds that "experiences with machines themselves are diverse and are not simply reducible to any single primitive set" (Ibid. 4). In his analysis of tool use in *Being and Time*, Heidegger shows that tools withdraw in our relationship with them and only become visible as concrete entities that have certain properties in instances of breakdown. Ihde follows Heidegger on this point but sets it as his task to uncover the variety of relations that one may have with technologies, and in which the world is constituted in a variety of ways.[3] In doing this, not the technology itself but how it makes us relate to the world becomes the center of our attention. In this respect, Ihde argues that Heidegger's phenomenological analysis of tool use can be extended to technologies.

The Ready-to-hand and Human-Technology Relations

Ihde attempts to capture the variety of ways in which technologies mediate our understanding and experiences of the world in a set of schemas indicating how technologies make us relate to the world. These can best be understood as specifications of the more general framework of the ready-to-hand that we find in Heidegger's work. However, at the same time, Ihde significantly expands on Heidegger by anticipating that structural critiques of Technology "become possible precisely by either an implicit or explicit choice of one type of relation as primary or paradigmatic" (Ibid. 4). In this sense, if Heidegger's phenomenology of tool use is interpreted as capturing *all* human-technology relations, it would be deprived of its analytic potential. This is why it is pivotal to understand technologies as *mediating* how human beings relate to the world, and to make clear that technologies do so by shaping intentional relationships between human beings in a variety of ways.

The most elaborate discussion of these relations can be found in Ihde's *Technology and the Lifeworld* (1990). He discerns four basic sets of relations with technologies: embodiment relations, hermeneutic relations, alterity relations, and background relations (Ihde 1990, 72–122). In each of these relations, technologies mediate between human beings and the world in a different way (i.e., make human beings experience and interpret the world differently).[4] Different technologies mediate how human beings are intentionally directed at the world, thereby opening and closing possibilities of experience and interpretation. This is what Ihde calls the *amplification-reduction* structure of technological mediations. By mediating the intentional relation between human being and world, there is an "amplification [that] tends to stand out, to be dramatic, while the reduction tends to be overlooked, or may be forgotten" (Ihde 1979, 21). In this sense, technological mediations are never neutral: they transform how human beings can relate to the world. Let me illustrate this claim in more detail by examining the first two relations Ihde discerned: the embodiment relation and the hermeneutic relation.

In *embodiment relations*, technologies mediate how our bodily experience of the world without becoming thematically present as an object in the world. In other words, in one's experience, one does not differentiate between what is one's own body and what is a technological extension of it; the body has incorporated the technological extension as being a part of it (Ihde 2010, 42). This can be illustrated by our experience of slicing bread. When I tear bread into smaller pieces with my bare hands, I may feel its warmth, softness, and so on. But when I use a knife to cut it, I experience the surface of the bread through the knife, that is, the carvings of the knife allow to cut the bread more easily and precisely (amplification), but I no longer experience the warmth of the bread (reduction). Furthermore, I do not experience the presence of the knife specifically; it makes me relate to the bread, not to the knife. In this example, the knife is incorporated in my bodily experience, making it impossible to distinguish between my body and the knife in this experience. In my relationship with the bread, the knife becomes *transparent*: it is no longer experienced as a knife but makes us experience the bread in a different way. This example shows the continuity between Heidegger's phenomenology of tool use and Ihde's postphenomenological framework because the technology *withdraws* in how it constitutes a relation with the world.

A second type of relation with the world mediated by a technology Ihde calls the *hermeneutic relation*. In this type of relation, one relates to the world through an artifact, but it does not become transparent. Contrary to an embodiment relation, one does not incorporate a technology into one's body schema, but the technology presents an aspect of the world in a specific way, such that it can be related to as an object of interpretation. Crucially, the technology already partly interprets the world for one; one's intentionality is directed at the technologically mediated phenomenon that enters our sensory space. The standard example that Ihde uses is the thermometer (e.g., Ihde 2010, 43). When relating to a thermometer, temperature becomes readable for me in a particular way. In this case, rather than being a tactile sensation that may induce feelings of warmth or cold, temperature becomes something that can be numerically appreciated in my relationship with a thermometer with a certain degree of precision. In this case, a thermometer constitutes a relation with the world that makes me understand temperature in a specific way. When relating to temperature through a thermometer, the bodily sensation of warmth is lost (reduction). However, it also allows for specific aspects to stand out, namely, the ability to specifically compare temperature over time in a precise way (amplification).

Also, in this relation, the thermometer itself does not appear to me as something that is present-at-hand; rather, it makes me relate to something else. It is something that I relate to in order to be able to know the specific degree of Celsius precisely. Just as in the case of embodiment relations, the thermometer withdraws in how it mediates my relationship with the world. In other words, it does not appear as foreign object but is integrated in how I interpret the world. While in terms of withdrawal, this relation is continuous with Heidegger's phenomenology of tool use, it augments it in terms of the different ways in which different technologies mediate my relationship with the world. Not only do I relate to technologies to discover the objects around me, but these also mediate how existing objects (such as temperature) can be interpreted. Understood in this way, technologies not only transform my capacity of *bodily* experiencing the world but also create new possibilities to appreciate the world *hermeneutically*.[5]

Disclosing the World through Technologies

How technologies mediate interpretations of the world is a theme further developed in Verbeek's *What Things Do* (2005). Central in his discussion of Ihde's human-technology relations is the idea that technologies do not merely allow one to relate to an already existing world in a variety of ways, but shape what one takes to be the world in the first place. In other words, technologies are not merely constitutive of how one interprets the world (as if they would allow establishing a relationship with an already existing set of entities) but are constitutive of the world proper:

> Of necessity, any description of reality cannot avoid being a rationale, explanation, or constitution. That is not to say that the world is only a construction, just that we cån never know the world as it is in itself, *but only as we disclose it.* An uninterpreted world, a world in itself, cannot be experienced; an untouched world cannot be lived in. Human beings never encounter a world in itself, only and always a world for them. (Verbeek 2005, 107, my emphasis)

In emphasizing that the world is disclosed within human-technology relations and that it does not make sense to speak about a world in another manner, Verbeek further radicalizes the notion of technological mediation: because intentionality is always the function of how one relates to a technology, the world is disclosed *through* one's relation with technology as well. In the mediated relation between human beings and technologies, "a specific 'objectivity' of world and a specific 'subjectivity' of human beings" is constituted (Ibid. 112). Accordingly, technological mediations should not be understood as *re*-presenting possible experiences and interpretations that already potentially resided in the world—or in the subject—but instead as *presenting* those, precisely because it only makes senses to speak about "the world" insofar it is constituted within human-technology relations.

Verbeek's use of mediation comes close to how Heidegger uses the notion of "worldhood": technologies are constitutive of the existential structure of human beings because they help in constituting both subject and world. Whereas Ihde takes Heidegger's analysis of tool use as a starting point for understanding our *relations* with the world, Verbeek seems to be closer to Heidegger, because it is within these relations that the notion of "world" attains its meaning. The world is a function of worldhood that takes shape in our everyday interaction with technologies. From this perspective, humans do not just relate to the world through technologies, but how the world is disclosed is also the function of their relationship with a specific technology.

For example, my smartphone does not just allow me to relate to the world in which the number of contacts that I can reach immediately has increased suddenly, but the very appearance of these contacts as contacts that can be immediately reached changes how the world becomes present to me. Both the individual (subject) and the world encountered (object) are specifically shaped in our relationship with a technology, in this case, a smartphone. At the same time, I become a subject with the capability of contacting people in an easy way, and the world is disclosed such that it becomes a world in which others are present as easy to be contacted. Thus, a specific artifact allows for the specific constitution of the world (i.e., *my* world in my relationship with it). This makes immediately clear that a technology is not something between me and a world external to me but constitutes how the world is revealed as is my relation to it. In this more radical interpretation of Heidegger's phenomenology of tool use, the focus is less on the withdrawal of tools in our relationship with them (as was the case in Ihde's appropriation of Heidegger) but increasingly on the variety of ways of disclosing the world that are mediated within human-technology relations.

2.3 HEIDEGGER'S PESSIMISM: DISCLOSING THE WORLD IN *THE QUESTION CONCERNING TECHNOLOGY*

To get a better grasp of the notion of technological mediation, it is important to be precise about the non-neutrality of technologies that it departs from. As Verbeek argues, precisely because technologies mediate relationships between technologies and the world, they should not be considered mute things at our disposal that allow us to pursue our preset goals more effectively:

> Contrary to what many people intuitively think, [. . .] technologies are not simply neutral instruments that facilitate our existence. While fulfilling their function, technologies do much more: they give shape to what we do and how we experience the world. And in doing so they contribute actively to the ways we live our lives. (Verbeek 2011, 1)

A similar intuition on the non-neutrality of technology can be found in Heidegger's famous essay "The Question Concerning Technology," where he writes that "we are delivered over to [technology] in the worst possible way when we regard it as something neutral" (Heidegger 1977a, 4). Through an analysis of Heidegger's seminal essay, I sketch the contours of the sort of non-neutrality of technologies Verbeek and Ihde have in mind when using the term "technological mediation."[6] However, as the next section demonstrates, how the non-neutrality of technology is appropriated within postphenomenology does not necessary amount to the conclusions about technology that Heidegger draws.

Technology's Non-neutrality

From the start of the essay, Heidegger's goal is clear: "We shall be questioning concerning technology, and in so doing we should like to prepare a free relationship to it. The relationship will be free if it opens our human existence to the essence of technology" (Ibid. 3). But what is this essence to which human existence must open up? To answer this question, Heidegger first identifies two common, interrelated conceptions of technology that do not reveal its essence: technology as a means to an end and technology as a human activity. He derives the following common definition of technology from these two conceptions:

> The manufacture and utilization of equipment, tools, and machines, the manufactured and used things themselves, and the needs and ends that they serve, all belong to what technology is. The whole complex of these contrivances is technology. Technology itself is a contrivance, or, in Latin, an *instrumentum*. (Ibid. 4–5)

The whole network of man-made tools and machines that are used to fulfill certain ends constitute a complex network that can be called "technology." From this perspective, technology is a complex instrument developed by human beings who have manufactured and connected tools and machines. In this definition, the damaging potential of technology lies in the fact that—with the advent of modern technology—it will fall outside of human control. From this perspective, establishing a free relation with technology depends on "'getting' technology 'spiritually at hand'" (Ibid. 5).

However, Heidegger holds that conceiving of technology in this manner is unable to reveal its essence, since "the essence of technology is by no means anything technological" (Ibid. 4). The essence of technology cannot be found when it is considered in terms of the tools and machines that can be described as "technological." While the above instrumental definition of technology is *correct*, it is not *true*, because it fails to reveal the essence of technology. However, this correct definition can function as a starting point for opening up to the essence of technology: "In order that we may arrive at this, or at least come close to it, we must seek the true by way of the correct" (Ibid. 6). Thus, when grasping why technology is commonly understood in terms of instrumentality and neutrality, it is possible to develop an understanding of its essence. This assumes that the common understanding of technology is not a "necessary" one but is instead a specific way in which technology is already disclosed. It is by analyzing how technology is disclosed in terms of instrumentality that its essence can be revealed.

An instrumental definition of the use of technology could be something like this: A uses B as a means to cause C, such that without any B, A would be unable to cause C. Consequently, there is a close connection between the notion of instrumentality and the notion of causality. But what form of causality is implied here? According to Heidegger, the modern way of understanding causality reduces it to the *causa efficiens*, which is only a specific way of causality when interpreted against the background of the different forms of causality distinguished by Aristotle. In modern times, this form of causality sets the standard for all forms of causality, thereby reducing it to the structure "B is effected by A." However, Heidegger holds that the word "cause" can be traced to the Greek *aition*. This etymological exercise reveals a profound other meaning of causality: as "that to which something else is indebted" (Ibid. 7). Contrary to the above characterization of the *causa efficiens*, this conception of causality is of the structure "B is indebted to A," thereby not interpreting the effect "B" to be identified as caused by a singular actor "A." Rather, this understanding opens up the possibility that B is indebted to something else.

But how then must we understand the manifestation of "B" as indebted to something else? Must we move back to the Aristotelian scheme and say that "B is indebted to A, C, D, and E," where A is the *causa efficiens*, C the *causa materialis*, D the *causa finalis*, and E the *causa formalis*? According to Heidegger, this would not point in the direction of the meaning of the Greek word *aition*, because in Aristotle's analysis of causality, it also remains unclear "whence does it come that the causal *character* of the four causes is so unifiedly determined that they belong together" (Ibid. 7) (Emphasis in original). What unites these four causes is that they are at play within what Heidegger calls *bringing-forth*. Being indebted in terms of *aition* is the bringing-forth of the concealed into unconcealment: Aristotle's four ways of causation are "unifiedly ruled over by a bringing that brings what presences into appearance" (Ibid. 10). In this sense, what unifies the four causes is a *poiesis*; it is an event in which something comes into appearance as this or that something. In this specific way, what characterizes what nowadays is called "causation" is the unconcealment of a thing in a specific way which is revealed (*entbergen*). Heidegger also proposes asking for the essence of technology in these terms: "What has the essence of technology to do with revealing? The answer: everything" (Ibid. 12). And since how technology is revealed is in terms of instrumentality, it is in this specific way of revealing that essence of technology can be opened up to.

The Essence of Modern Technology: The Enframing of Nature

But in what sense does how technology reveals itself as instrumental simultaneously reveals its non-neutrality? To explore this question, Heidegger illustrates why Technology must be understood as a way of revealing by discussing the bringing-forth of the Rhine through the presence of a hydroelectric plant:

> The hydroelectric plant is set into the current of the Rhine. It sets the Rhine to supplying it hydraulic pressure, which then sets the turbines turning. This turning sets those machines in motion whose thrust sets going the electric current for which the long-distance power station and its network of cables are set up to dispatch electricity. In the context of the interlocking processes pertaining to the orderly disposition of electrical energy, *even the Rhine itself appears as something at our command.* (Ibid. 16, my emphasis)

It is in this respect that Heidegger understands technology as non-neutral: through the presencing of the Rhine by the hydroelectric plant, the Rhine is revealed in terms of instrumentality (e.g., as a source of electricity). Technology manifests its non-neutrality by bringing the Rhine out of concealment as something at our command rather than revealing it as something that inspires awe and admiration.

While it is possible to interpret this as a desirable development—one must only be remembered of the Baconian ideal of mastering nature—Heidegger conceived of this way of revealing as a *challenging* of nature: It "puts to nature the unreasonable demand that it supply energy that can be extracted and stored as such" (Ibid. 14). This mode of revealing makes it impossible to interpret the Rhine as a source of artistic inspiration but only reveals it as a source of calculable energy.[7] When challenging nature to reveal itself in terms of calculability, technology as a way of revealing transcends the particularities of technologies. In fact, Heidegger argues that also the development of modern physics should be understood as a technological way of disclosing the world: "Modern physics, as experimental, is dependent upon technical apparatus and upon progress in the building of apparatus" (Ibid. 14). As a way of revealing, Technology is a worldview that challenges nature forth as being unconcealed in terms of calculability and manipulability.[8] This is what Heidegger calls the ordering of nature as a *standing-reserve* (*bestand*) (Ibid. 17). According to Heidegger, it is through this particular way of bringing-forth as challenging-forth that nature is revealed as something at human disposal, as is the Rhine as brought into concealment through the presence of the hydroelectric plant.

However, since it was mankind that was responsible for the installation of a hydroelectric plant, does this not presuppose that the challenging-forth of nature is caused by mankind rather than by technology? Anticipating this criticism, Heidegger shows that such a perspective assumes that there is something that can be called "mankind" prior to Technology. Yet, mankind participates in how the unconcealed is revealed through Technology: "The unconcealment of the unconcealed has already come to pass whenever it calls man forth into the modes of revealing allotted to him" (Ibid. 19). Before man sets a technology such as an energy plant into the current of the Rhine, nature is already concealed as a standing-reserve (i.e., as something that is at our command). This is what Heidegger calls the Enframing (*Gestell*) of nature, which "concentrates man upon ordering the real as standing-reserve" (Ibid. 19). In this sense, the essence of Technology must be understood in terms of non-neutrality: if inevitably discloses of nature as standing-reserve. The essence of technology as Enframing is not a passive property of Technology but is conceived as something active: it unconceals nature from concealment as standing-reserve. The uniformity of the standing-reserve makes it impossible, through how other appropriations of nature are ruled out when revealed through Technology, to relate to nature in other possible ways. It is only in terms of instrumentality that it can be appreciated.

Technology's Neutrality as Threat

Does the Enframing leave us with no choice than being delivered over to Technology, or does it allow for developing new relations with nature? Heidegger is ambivalent on this point: "When we once open ourselves expressly to the *essence* of technology, we find ourselves unexpectedly taken into a freeing claim" (Ibid. 26) (Emphasis in original). However, this "freeing" has nothing to do with the bringing under control of a technological system that has got "out of our hands." This would assume that, in the end, the essence of technology is something neutral that can be brought under our control. When dismissing this perspective, it becomes clear what Heidegger really is criticizing:

> The treat to man does not come in the first instance from the potentially lethal machines and apparatus of technology. The actual threat has already affected man in his essence. The rule of Enframing threatens men with the possibility that it could be denied to him to enter into a more original revealing and hence to experience the call of a more primal truth. (Ibid. 28)

When interpreted as something neutral, man is incapable of seeing that Technology challenges nature as standing-reserve and is prior to every human involvement. This is why he stresses at the beginning of the text that the worst way to conceive of technology is to conceive of it as something neutral: this would rob us of the possibility to open us to technology's essence.

What is at stake for Heidegger is that not being opened up to the essence of Technology prevents man from understanding their own way of being (i.e., how the unconcealed is brought out of concealment by the Enframing). What Heidegger eventually calls for is an opening up to the essence of technology of the Enframing, and he does not present it is a "problem" that demands a "solution"—the very idea of a "solution" to the Enframing is absurd from a Heideggerian perspective. Instead, it is the recognition that Technology is a way of revealing that cannot but unconceal nature as standing-reserve that might constitute a relation between humans and how they *are*. Cautiously, one can say—invoking the terms that Heidegger uses in *Being and Time*—that Technology reveals the "worldhood of the world," and thereby mediates how human beings *are* in this world.

2.4 ENFRAMING AND MEDIATION IN HUMAN-TECHNOLOGY RELATIONS

Heidegger hardly discusses any concrete technological practices in "The Question Concerning Technology" because his starting point is that the *essence* of technology is "by no means anything technological" (Heidegger 1977a, 4). As a consequence, any investigation into the nature or working of individual technologies would by definition not be of interest to his particular philosophical concern. Postphenomenologists retain Heidegger's idea that technologies are non-neutral in the sense that they are co-constitutive of human-technology relations, thereby mediating how human beings relate to the world. However, they criticize Heidegger's later philosophy of technology because it presupposes a uniformity of "the technological" that transcends the practices in which technologies partake (Verbeek 2005, 80, cf. Zwier et al. 2016). However, to transform the observation that Heidegger did not pay attention to the workings of individual technologies into a criticism against him, it is necessary either to show the relevance of individual artifacts in the search for the essence of technology or to show that the philosophical question regarding the essence of technology is misguided. In other words, it must be shown that the essence of technology is by any means technological, or that Technology has no uniform essence.

Postphenomenology's Criticism of Transcendentalism

As noted by Peter-Paul Verbeek, the portrayal of Heidegger's philosophy of technology as too abstract and monolithic is *gratuit* when based on an external assessment of Heidegger's philosophy of technology, without engaging with Heidegger on his own terms (Verbeek 2005, 48). More fine-grained arguments against Heidegger's philosophy can be found within postphenomenology and are explicitly formulated in Verbeek's *What Things Do* and in Don Ihde's *Technics and Praxis.*

The central element of the postphenomenological criticism against Heidegger's philosophy of technology is that it confuses the workings of technologies with their conditions of possibility (i.e., Enframing) and therefore fails to notice the several different ways in which technologies disclose the world (e.g., Ihde 1979, 106; Verbeek 2005, 75). Heidegger and postphenomenologists share a critique of any instrumentalist understanding of technology that conceives of it as a neutral means. However, postphenomenological thinking breaks with Heidegger's analysis in its rejection of the uniform manner of revealing nature that Heidegger attributes to Technology.

The primary criticism of postphenomenology against Heidegger's position is the idea that it is possible to strictly differentiate between the workings of technologies and their conditions of possibility. As such, the difference between the two positions is grounded in a different definition of "technology." Whereas Heidegger from the start assumes that Technology has a uniform "essence" that can be opened up to, postphenomenologists maintain the view that the actual practices in which technologies mediate relationships between human beings and the world are so diverse that any analysis of technology should reflect this diversity. *If* one would like to speak about technologies in terms of essence(s), this should be done through an analysis of technological practices, and the different ways in which technologies manifest in practices reveal that their essence is in constant flux—which in a sense make any attempt to search for an essence obsolete. Therefore, rather than understanding Technology in terms of uniformity (i.e., assuming that Technology is a structural phenomenon that can be approached in a uniform manner), a postphenomenological perspective maintains that a philosophy of technology should be grounded in the analysis of concrete technologies (e.g., Ihde 2010, 18).

To use analyses of concrete technologies as a criticism of Heidegger, it is necessary to show that the essence of technology *is* something technological. Only as such can the variety of ways in which the world is disclosed through technologies function as an argument against the philosophical ambition to ground technology in a *singular* essence. According to Verbeek, Heidegger's different treatment of concrete technologies and his preference for premodern technologies already hint at the idea that the essence of technology is not something that is "above" the things:

> Without the selectivity in Heidegger's approach either the hydroelectric plant would have to reveal the fourfold as well or the bridge could not be understood as a thing that essentially (ahistorically) reveals the event of coming into being. In other words, Heidegger's understanding of modern technology in terms of

alienation, in which something originary gets lost, would not have been possible without applying a double standard (Verbeek 2005, 74).

As Verbeek argues, there is nothing wrong with nostalgia per se: after all, why should we accept a priori that we should not prefer premodern technologies over modern ones (Ibid. 68)? However, revealing Heidegger's nostalgia as a consequence of his investigation of the conditions of possibility of technologies allows its transformation into a criticism. From this perspective, Heidegger arbitrarily attributes the Enframing to the hydroelectric plant and, on this ground, distinguishes it from the (non-Technological) disclosing of the Rhine by a bridge. While it is perfectly possible that the bridge and the hydroelectric plant disclose the Rhine in a different manner, there is no fundamental reason why one should be preferred over the other. It is precisely this difference that Verbeek considers irreducible to the conditions of the possibility of Technology; on the contrary, it is this difference that demands philosophical analysis. This does not necessarily imply that investigations into the conditions of possibility of technologies are unable to offer insights into the workings of technology, but it shows that Heidegger's transcendental interpretation of technology cannot account for the operativity of technologies in society and hence is unable to acknowledge that technologies can constitute new practices in society and new interpretations of reality (Ibid. 9).

2.5 CONCLUSION: TECHNOLOGICAL MEDIATION THEORY AS A NON-TRANSCENDENTALIST APPROACH TO NON-NEUTRALITY

Although Heidegger surely acknowledges the non-neutrality of technologies—actually, this is central in his idea that they challenge nature as standing-reserve—his transcendentalist understanding of non-neutrality fails to bring to light the different ways of being that technologies help constitute. For example, he sees the revealing of the Rhine as standing-reserve by the hydroelectric plant as the only way that this specific technology discloses the world, thereby neglecting that it also allows families to light their houses or lets hospitals function more efficiently. As Ihde shows, Heidegger's negative evaluation of the typewriter has a similar connotation. Heidegger writes, for example, that "the typewriter tears writing from the essential realm of the hand—i.e., the realm of the word" (Heidegger 1992, 119). Sure, there are relevant differences between handwriting and writing on a typewriter, and this difference is worthy of investigation. However, on what ground can we a priori determine that handwriting allows for entering the essential realm of the word, while this possibility is removed by the typewriter (Ihde 2010, 77)?

When extending this example to the contemporary use of the computer, the lack of explanatory value of Heidegger's transcendentalist perspective becomes even clearer: even when accepting that the essential realm of the world is concealed through the computer, must one not also attempt to understand how computers and similar devices enable us to communicate more quickly and how the use of computers and more developed internet and communication technologies (ICT) allows for a fundamentally different way of understanding—and dealing with—cases of natural disasters, to give an example?

The lack of explanatory value in Heidegger's conception of technology is not necessarily the consequence of his understanding of technology in terms of its conditions of possibility, but is rather caused by his distinguishing between the actual workings technologies, on the one hand, and their conditions of possibility, on the other. In postphenomenology, these two appear as intertwined, which makes it possible to make both the different workings of technologies and how they disclose the world subject to philosophical analysis. This does not necessarily prevent one from understanding the hydroelectric plant as challenging forth the Rhine as a standing-reserve (i.e., as a mere source of energy). However, it shows that how the world is revealed through the hydroelectric plants cannot be reduced to this specific event. Furthermore, how electricity functions in society cannot be traced back to the place where it was produced, as its concrete manifestations differ in different times and periods. For instance, the hydroelectric plant gives rise to new technological developments that were not incorporated in its original challenging of nature. This is not to say that there are essential features of the hydroelectric plant that should be preferred over the features of a bridge over the Rhine. After all, electricity has also been key in the production of advanced weaponry that might cause great damage. It is in this sense that technological objects should be interpreted as revealing the world within their specific historical embedding; it is only in this aspect that how reality is revealed in relation to them can be appreciated.

The Heideggerian understanding of the essence of technology as a way of revealing leads to an ahistorical understanding of technology. Neither humans nor technologies are capable of relating to this typical way of revealing consciously or unconsciously. This is most clearly formulated in the idea that technology eventually prevents man from understanding his own essence: "In truth, however, precisely nowhere does man today any longer encounter himself, i.e., his essence" (Heidegger 1977a, 27). This assumes that the essence of technology is preventing access to an authentic human essence that is independent of time and place. In both cases, an essence is not something that is found *in* concrete reality but rather is an ahistorical concept existing outside space and time. On this point do postphenomenology and

Heidegger diverge most clearly: while Heidegger interprets essence as something that is universal, postphenomenologists stress that it is in constant flux because it changes within specific practices. To illustrate this shift, postphenomenologists speak of technologies as *mediating* relations between human beings and the world, rather than conceptualizing them as a way of *revealing* (*entbergen*). The concept of mediation reveals that because technology is a dispersed phenomenon scattered across different individual technologies, it opens up multiple ways to relate to the world. Accordingly, the "essence" of technology is in constant flux, as is our understanding of reality.

In concrete practical environments, technologies disclose the world in a specific way. However, this disclosure does not occur in isolation, independent of the presence of other technologies and human beings. When one must think of technologies as embedded in concrete practices, it becomes clear that they disclose a variety of realities in a variety of ways. Conceptions of technology as alienating human beings from their own authenticity or as determining the "structure" of modern society do not appear as adequate within this postphenomenological perspective. Surely, "alienation" can be one manner in which a particular technology reveals reality in a concrete practice. However, several other ways of disclosure are possible as well: just as technologies can instantiate feelings of alienation, they can create new experiences that could not be imagined without them, such as advances in medical care, which induces a fundamentally different understanding of health and hygiene. The crucial point is that regardless of those being positive and negative, technologies constitute subjectivity and objectivity in a specific way, thereby making possible new ways to understand and experience the world. To capture this diversity, postphenomenology uses the term "technological mediation": this term is continuous with Heidegger's later work, which sees technologies as disclosing the world, but simultaneously breaks with it because it takes Heidegger's phenomenology of tool use as a starting point and sees the disclosing of the world as an event that occurs *within* the relation between human beings and technologies. In this book, I build on this understanding of the notion of technological mediation to understand how the reality that scientists study is revealed to them in a particular way when relating to particular technologies.

NOTES

1. Although the German word *Zeug* is translated as "equipment" in Macquarrie and Robinson's 1985 translation of *Being and Time*, I use the term "tool" to be consistent with the way *Zeug* is usually translated within the philosophy of technology.

2. Of course, if I wish to repair this computer, the specific workings of it may become relevant. However, from Heidegger's perspective, this would imply that my

computer appears in a different totality. Rather than the computer itself, the tools I use to repair the hardware or software of the computer are the things that I primarily relate to against the larger background of repairing the computer.

3. Note that "world" in this context is used in a Heideggerian sense as discussed in the previous section. Thus, rather than referring a set of objects external to us, it is interpreted as constituted through worldhood.

4. A detailed discussion of the specific structure of these relations can be found in Verbeek (2005, 123–28).

5. Herewith, I do not mean to imply that there is a strict difference between having an embodied and a cognitive (or interpretative) relation with the world. However, the structure of a *hermeneutic* relation is one in which primarily interpretative capacities are mediated by a technology.

6. I do not aim to give a comprehensive discussion of *The Question Concerning Technology* and its place in Heidegger's oeuvre, but I focus on the differences and similarities in how the non-neutrality of technology is discussed by Heidegger and within postphenomenology. For a detailed general discussion of *The Question Concerning Technology*, see, for example, Loscerbo (1981, 129–53) and Visser (2014).

7. Anticipating the postphenomenological discussion of Heidegger's work, it should be noted that ordering the Rhine as a source of electricity and as instantiating feelings of awe and admiration are not necessarily mutually exclusive.

8. This is the central claim of Heidegger in his essay "The Age of the World Picture," which discusses the conditions of possibility of modern science, and is further discussed in the next chapter that specifically deals with Heidegger's evaluation of modern science and its relation to modern technology.

Chapter 3

Science and the Theoretical Disclosure of Nature

INTRODUCTION: SCIENCE AS PRACTICE?

When a scientist arrives at a laboratory, (s)he encounters a technological infrastructure that is only comprehensible for people who are trained to use it. Contrary to the use of my coffee machine, it requires a particular theoretical background and extensive training to be able to engage with this specific set of technologies. For example, when a cognitive neuroscientist arrives at work and is finished answering her e-mails, (s)he may engage in data analysis through the use of highly specialist calculation programs, or she may test a participant in an MRI scanner, of which the outcome will become part of a data set that must be analyzed. On the basis of a specific theoretical background, then, (s)he might end up interpreting a brain scan by linking its content to a particular aspect of human cognition. It is clear that this scientist is equally unable to function without the presence of a large technological infrastructure. Considering that both our scientific and daily practices crucially depend on technologies, can we still distinguish between the two, or must we acknowledge that both depend on technology in a similar way? In this chapter, this question is explored by asking to what extent postphenomenological analyses of human-technology relations as introduced in the previous chapter can be fruitfully integrated into a philosophy of scientific practice.

Within the field of postphenomenology and the related field of STS, the dependency of both scientific and ordinary practices on technologies resulted in the idea that in one way or another, a continuum exists between the two (e.g., Alač 2008; Ihde 1991, 1998; Rosenberger 2008; Verbeek 2005, 2011; Vertesi 2012). In all of these cases, it is tacitly assumed that the framework used to analyze technologies in everyday life can be straightforwardly extrapolated to the domain of scientific practice. However, is it warranted to do so, or is there something distinct about science that instantiates a specific use of technologies? For example, is the dependency of scientific research on existing scientific theories—something that is often absent in our daily practice—reflected in how technologies are used in scientific practice? Or, more generally, is the reality to which scientists relate essentially the same as the reality encountered in everyday life?

In this chapter, these questions are addressed by exploring whether the postphenomenological criticism on Heidegger's analysis of Technology as being overly monolithic that culminates in a philosophy of technological mediation can be unproblematically extended to Heidegger's analysis of science. The previous chapter showed that when focusing on individual technologies, nature need not appear as standing-reserve. In the postphenomenological literature, Heidegger's analysis of science is dismissed along similar lines, with reference to the idea that if Technology is ontologically prior to science, Heidegger is committed to the idea that the essence of science can be traced back to the ordering of nature as standing-reserve (e.g., Ihde 2010). This seems to suggest that Heidegger's thinking about science should be understood as analogous with his thinking about Technology. However, as I show in this chapter, in Heidegger's thinking, a starting point for developing a *hermeneutics* of scientific practice can be found that attempts to capture both the *practical* and the distinct *scientific* character of scientific practice. I do so by showing how postphenomenological analyses of the role of technologies in scientific practice neglect an important insight of Heidegger's hermeneutics of science: namely, how in scientific practices, a relation with objects appearing as present-at-hand is established.

3.1 HEIDEGGER AND THE PRIMACY OF PRACTICE IN SCIENCE

In the philosophy of technology, Heidegger's thinking about science is often discussed on par with his philosophy of technology, and critics dismiss it as being unable to account for positive developments in science and medicine (e.g., Feenberg 1999; Ihde 2010; Svenaeus 2013). While this may be partly true for his questioning for the essence of science in his later works, Heidegger's earlier thinking was not dismissive of science per se, only insofar as it claimed to be ontologically foundational (Caputo 2012). In *Being and Time*, Heidegger was concerned with developing a hermeneutics of science that understood science as a specific kind of practice explicitly concerned with disclosing the world as a set of present-at-hand objects (Glazebrook 2000b; McManus 2012). In this context, Heidegger asks how the possibility to discover the world in terms of present-at-hand objects can arise, given that our initial contact with the world always occurs in relation with the ready-to-hand.

Rather than understanding science in terms of a set of theoretical propositions in need of justification, Heidegger's hermeneutics of science takes science to be a "way of existence and thus as a mode of being-in-the-world" (Heidegger 1985, 408; see Nordmann 2012), in which the world and the entities in it are disclosed in a specific way. As a way of existence, science is being constituted through the existential structure of human beings, and is, therefore, in Heidegger's view, a fundamentally worldly activity that cannot be separated from the human being that engages in it. Because of this, "nature [is revealed as] an entity which is encountered within the world and which can be discovered in various ways and at various stages" (Heidegger 1985, 64). According to Heidegger, precisely because our initial contact with the world is not one in which it appears as a set of objects with certain properties, a specific way of disclosing the world is presupposed when engaging in scientific research.

It is important to note that Heidegger does not dismiss that the sciences may be very informative and yield genuine results; what is at stake is instead that science can never function as the basis of an ontology, as it is derivative of a specific disclosure of the world rather than capturing the most accurate way to disclose it. Heidegger calls this the *projection* of the world in which it becomes present as a set of objects that are present-at-hand and can be made subject to theoretical investigation (Glazebrook 2000b, 5). In other words, disclosing the world as a set of objects appearing as present-at-hand is a specific mode of disclosure that is dependent on a specific kind of practice.

Heidegger points to the fact that the scientific disclosure of the world relies on practical interactions with material technologies: "Theoretical research is not without a *praxis* of its own. Reading off the measurements which result from an experiment often requires a complicated 'technical' set-up for the experimental design. Observation with a microscope is dependent upon the production of 'preparations'" (1985, 409). In Denis McManus's book *Heidegger and the Measure of Truth*, this is summarized as Heidegger's *Primacy of Scientific Practice* thesis, which indicates that "scientific knowledge presupposes—is bound up with, is unintelligible without reference to—certain practices" (2012, 63). Thus, in one way or another—and this is what commentators on Heidegger generally agree upon—"the ready-to-hand is prior to, or more fundamental than, the present-at-hand" (Ibid. 69).

How then must the priority or more fundamental nature of the ready-to-hand in this context be understood? It seems that from Heidegger's analysis of worldhood, it follows that the priority of the ready-to-hand can be found in how it is constitutive of the worldhood of Dasein, and thereby prior to a structural whole of "world." As such, it is clearly not a thesis about the priority of ready-to-hand to present-at-hand in the sense that the latter can be reduced to the former. On the contrary, the "ready-to-hand" and the "present-at-hand" do not exist independently of human beings but take shape in the relation between human beings and the world. Heidegger's thesis concerns the relationship between human beings and the world as shaped by the relation with the things around them. Accordingly, both the ready-to-hand and the present-at-hand must be understood as *modes* through which a relationship with reality is established. Later in this chapter, I draw on postphenomenology to show how the ready-to-hand is considered prior to the present-at-hand in the constitution of these modes.

An Existential Conception of Science: Deliberately Relating to Reality as Present-at-Hand

Heidegger discusses the transformation of a relation with the ready-to-hand into a relation with the present-at-hand in §69b of *Being and Time*, in which he develops an "*existential conception of science*" (1985, 408) (Emphasis in original). Whereas he initially discussed the present-at-hand as emerging in occasions of breakdown (Ibid. §§16–18), in this section, he considers how this disclosure is constituted *deliberately*. Constituted in this way, entities of concern are brought closer to the investigating subject, such that they can become objects of study to which a certain set of properties can be described. This is realized through stepping over our initial contact with the world (i.e., the ready-to-hand character in which entities become normally present to us must be turned over). Hence, it is not only because of a breakdown that it becomes possible to relate to objects as being present-at-hand; the deliberate disclosure of the world in this way is not suddenly caused by accident.

According to Heidegger, the primary concern of science is to make the totality of ready-to-hand present as if the individual entities in this totality do not relate to one another, but instead to make aspects of specific entities explicitly stand out. This does not imply that the totality of a world structured in relation with the ready-to-hand is overcome, but rather that a different totality is disclosed. "The scheme peculiar to this [totality] is the 'if-then'; if this or that, for instance is to be produced, put to use, or averted, then some ways and means, circumstances, or opportunities will be needed" (Ibid, 410). The very possibility to relate to an entity as present-at-hand, the "then" in the above fragment, presupposes an "if," making it necessarily dependent on a ready-to-hand encounter of the world.

It is thus through particular practical concerns that a situation in which objects appear as isolated from the whole of which they are a part can manifest. To illustrate this, Heidegger returns to the example of the hammer introduced in §15 of *Being and Time* and states there are two different ways of relating it. On the one hand, the hammer can be assessed in terms of whether it is *too* heavy to perform a certain job; in doing this, it becomes present-at-hand because it no longer withdraws. On the other hand, the hammer can be related to, being part of a different totality, as an object that has the property of heaviness that can be exactly measured. In this way, the hammer is cut loose of its initial environment, such that it is dissociated from the practice of hammering. Therefore, the hammer can no longer be assessed in terms of hammering, which makes it no longer sensible to speak of heaviness in terms of "too heavy." Heidegger proposes interpreting this as a modification of the space in which a hammer is encountered: "Its place becomes a matter of indifference. This does not mean that what is present-at-hand loses its 'location' altogether. But its place becomes a spatio-temporal position, a 'world-point,' which is in no way distinguished from any other" (Ibid. 413).

In emphasizing the modification of the space in which an object is encountered, it becomes clear that to relate to a "physical nature" implies disclosing space in terms of a well-defined spatio-temporality, thereby leaving behind the totality in which physical objects are normally encountered. This is what Heidegger calls the mathematical disclosure of the world; relating to objects as present-at-hand is to project nature as mathematical. Only if nature is encountered in this way, "anything like a 'fact' [can] be found and set up for an experiment, regulated and delimited in terms of this projection" (Ibid. 414). Heidegger's hermeneutics of science (i.e., his hermeneutics of the mode in which a relation with the present-at-hand is established) has two important consequences that need to be accounted for when studying technological mediations in *scientific* practice: (i) for science to function, a mathematical disclosure of nature is required, and (ii) this mathematical disclosure is grounded in the ready-to-hand, that is, in scientific practice.

This clarifies that two different possibilities for relating to an object that is present-at-hand can be distinguished. On the one hand, objects can be present-at-hand through *breakdown*, as discussed in section 2.1, such that the relation to it is the consequence of the sudden breakdown of a tool against a certain background. On the other hand, Heidegger introduces an alternative mode of establishing a relation with the present-at-hand, discussed in this section, which I term the "Deliberative Mode."

3.2 SCIENCE AS RE-SEARCH

In Heidegger's later work, science is no longer a special way of disclosing the world *as* present-at-hand but is equated with the general technological disclosure of world as standing-reserve. Contrary to his earlier analysis in *Being and Time*, he no longer differentiates between the disclosure of the world in general and the disclosure of the world specific to scientific practice. As such, science becomes part of the modern epoch, in which nature is already disclosed as standing-reserve. However, through an analysis of "The Age of the World Picture" and "Science and Reflection," I attempt to show the ambiguity in Heidegger's later understanding of science. On the one hand, he criticizes the disclosure of the world through science as ruling out all other possible relations with it, because the essence of technology is presupposed in the science's way of revealing (e.g., Heidegger 1977a, 23). On the other hand, he seems to be uncritical of a scientific disclosure of the world per se.[1] The point of this detour is to show that it remains possible to adhere to the view that science presupposes a specific mode of relating to the world without monopolizing science as the only way of disclosing the world.

The difference between Heidegger's early hermeneutics of science and his later analysis of the essence of science is best exemplified by comparing a fragment from *Being and Time* with one from "The Question Concerning Technology." In the former, the scientific disclosure of the world is explicitly thematized in terms of the possibilities through which the world can appear *as* scientific: "Even the 'unity' of the manifold present-at-hand, of Nature, can be discovered only if a *possibility* of it has been disclosed" (Heidegger 1985, 184). In the latter, this "as-structure" is no longer understood as one possibility among others, but the scientific disclosure of the world is an omnipresent way of revealing: "When man, investigating, observing, ensnares nature as an area of his own conceiving, he has already been claimed by a way of revealing that challenges him to approach nature as an object of research" (Heidegger 1977a, 19). In this latter case, the world can no longer be disclosed *as* scientific, but scientific practice already participates in a particular way of revealing. This indicates that the ambiguity found in Heidegger's later work is grounded in that he no longer considers science as a special case of worldhood, but is part of a prior technological way of revealing. As such, Heidegger no longer conceives of having a mathematical relation to the world as particular to scientific practice, but instead as the only possible mode through which the world can be related to (cf. Glazebrook 2000a, 95).

The Essence of Modern Science

What exactly is this mathematical relation with nature? Heidegger explicitly connects it to a form of exactness particular to modern science. Exactness demands that "all events [. . .] must be defined beforehand as spatiotemporal magnitudes of motion" (Heidegger 1977b, 119). This presupposes that science has a particular "object sphere [that] has the character of exactitude" (Ibid. 120). According to Heidegger, the exactitude that is presupposed in modern science makes that the essence of modern science should be characterized as *research.* For Heidegger, this implies that the sciences project a pre-given space in which natural events happen and clarifies what types of objects can be observed in this domain.[2]

Because research is its essence, modern science must be distinguished from practices of knowledge gathering in ancient Greece or in the Middle Ages. Heidegger holds that it does not make sense to compare modern science with ancient Greek inquiries into nature, because "the Greek understanding of the essence of body and place and of the relation between the two rests upon a different interpretation of beings and hence conditions a correspondingly different kind of seeing and questioning of natural events" (Ibid. 117). The sphere of beings corresponding to the Greek way of questioning nature thus differs from how the world is projected through modern science.

Because modern science is essentially research and presupposes a mathematical way of disclosing the world, the sphere of objects it aims to describe is thought of as principally containing no riddles (i.e., in principle, everything in the world would be controllable or predictable when research with sufficiently rigorous exactitude is conducted). Heidegger attempts to clarify this with reference to the Greek understanding of mathematics:

> *Ta mathêmata* means for the Greeks that which man knows in advance in his observation of whatever is and in his intercourse with things: the corporeality of bodies, the vegetable character of plants, the animality of animals, the humanness of man. Alongside these, belonging also to that which is already-known, i.e., to the mathematical, are numbers. If we come upon three apples on the table, we recognize that there are three of them. But the number three, threeness, we already know. This means that number is something mathematical. (Ibid. 118–119)

The specific shape mathematics takes within modern physics is derivative of how nature *is* already revealed. As such, the worldness of the world, the question what and how the world is, has already been answered before any specific investigation of any natural phenomenon has started. This makes it so that it cannot be questioned *that* the world is a "self-contained system of motion of units of mass related spatiotemporally" (Ibid. 119). This is what characterizes the essence of science as research: the world must be constituted such that it can be approached in terms of a rigorous exactitude, such that all events in nature are conceptualized as spatiotemporal motions in a given domain.

Science and the Danger of Technological Projection

Given that Heidegger defines the essence of modern science as research and establishes a close connection between science and technology, how does his understanding of modern science as research relate to his idea that technology reveals nature as standing-reserve? Does the mathematical disclosure of the world through modern science by necessity have the same consequence? This issue becomes especially pressing when considering the reliance of science *on* technology. Technological procedures and machineries are necessarily a reflection of the exactness of the object sphere essential to modern science: "Within the complex of machinery that is necessary to physics in order to carry out the smashing of the atom lies hidden the whole of physics up to now" (Ibid. 124). It is only in this way that science remains capable of grasping nature in terms of its own projecting. According to Heidegger, the technologies through which the objects of the sciences are approached thus cannot transcend the boundaries of the object sphere that is defined through the research character of modern science. This does not rule out that through technological developments, a potentially infinite set of experiments can be developed through which new aspects of this sphere can be made visible. However, in Heidegger's view, this will not result in new forms of world constitution, since every experimental result is the function of the predefined domain that is approached. This is what he calls science as *ongoing activity* (*betrieb*).

Again, it can be asked if this must be considered as problematic. After all, developments in physics have allowed us to uncover the subatomic structure of the world and medical treatments have greatly increased due to a better understanding of cell structures. Does this not imply that the effectiveness of science is the consequence of the "research" character of modern science? Heidegger's ambivalent stance toward modern science is most clearly visible in "The Age of the World Picture." Let me first quote at length his primary concern with science as research:

> The scholar disappears. He is succeeded by the research man who is engaged in research projects. These, rather than the cultivating of erudition, lend to his work its atmosphere of incisiveness. The research man no longer needs a library at home. Moreover, he is constantly on the move. He negotiates at meetings and collects information at congresses. He contracts for commissions with publishers. The latter now determine along with him which books must be written. The research worker necessarily presses forward himself into the sphere characteristic of the technologist in the essential sense. (Ibid. 125)

In this fragment, it seems that the emergence of a predefined object sphere has the consequence of leading to a specific kind of individual that must be developed to work in such an environment. There is no longer room to think because the set of problems that must be handled is already defined. As a consequence, scientists run the risk of turning into businessmen who must fulfill the needs of external parties rather than spend time on science itself. If this were true, the distinction between scientific research and technological businesses would cease to exist. Accordingly, science should be understood on par with other parties with technological interests and would belong to a similar schematic, in which the ordering of nature as standing-reserve through science would be reinforced. However, in a footnote accompanying the above quote, Heidegger nuances his initial negativity:

> The phrase "ongoing activity" [*Betrieb*] is not intended here in a pejorative sense. But because research is, in essence, ongoing activity, the industrious activity of mere "busyness" [*des blossen Betriebs*], which is always possible, gives the impression of a higher reality behind which the burrowing activity proper to research is accomplished. Ongoing activity becomes mere busyness whenever [. . .] [it] never again confirms and verifies its own self-accumulating results and the calculation of them, but simply chases after such results and calculations. (Ibid. 138, note 2)

While the danger of becoming another branch of business always remains, science need not necessarily act as such. Science as research always faces the danger of being reduced to business, but when relying on its own critical potential, it still remains possible to reflect on its calculations and results. This suggests that modern science can become aware of the specific mode of disclosing the world in which it engages and hence come to terms with the domain of objects that it investigates being defined beforehand. Accordingly, there always remains a possibility to recognize the limits of the mathematical disclosure of the world and the contingencies involved in the particular set of objects under investigation.

But how can science enact this critical potential? And to what extent does the ubiquitous presence of technologies in scientific practice threaten science to become "mere busyness"? If the essence of science is ontologically conditioned by the essence of technology, then it seems that science must concede that its critical potential cannot be enacted. Thus far, Heidegger still leaves open the question of whether the sciences are victim to the same scheme as technology and must necessarily approach nature as standing-reserve. However, this is precisely the question at stake.

In "The Age of the World Picture," Heidegger in the end seems to interpret the culmination of the essence of science as research as part of a historical mode of revealing peculiar to Modernity (i.e., as being ontologically conditioned by a technological way of revealing). It is only in this epoch that the world is encountered as something to which man is opposed and is represented *by* man as opposed to him: "Therewith man sets himself up as the setting in which whatever is must henceforth set itself forth, must present itself, i.e., be picture. Man becomes the representative of that which is, in the sense of that which has the character of an object" (Ibid. 132). In other words, in Modernity, the world is always already disclosed as objectified *by* man and is disclosed *to* man as objectified consequently. Through the active engagement with nature that allows scientists to relate to the world as a set of objects, the ordering of nature as standing-reserve seems to be reestablished. From this perspective, the conditions in which science as research occurs simultaneously seem to constitute its limitations.

3.3 SCIENCE, TECHNOLOGY, AND HEIDEGGER'S AMBIGUOUS PESSIMISM

The question now arising is, *if* a lack of critical potential is present in scientific research, will it also force its particular way of disclosing the world onto other domains? In his essay "Science and Reflection," Heidegger answers this question affirmatively, because "the reality within which man of today moves and attempts to maintain himself is, with regard to its fundamental characteristics, determined on an increasing scale by and in conjunction with that which we call Western European science" (1977c, 156). Again, one might ask to what extent this would be a problem, given the positive developments accomplished through scientific research? In "Science and Reflection," Heidegger explicitly addresses this question by discussing the relationship between the form of world disclosure presupposed in modern science and the possibility to reflect on it. According to Heidegger, *if* a critical relation to the sciences is to be established, it must develop a form of reflection independent of the sciences, because the sciences are—as research—bound to the limits of a mathematical disclosure of reality.[3] This rules out the possibility of reflection, because "if we want to assert something about mathematics as theory, then we must leave behind the object-area of mathematics, together with mathematics' own way of representing. We can never discover through mathematical reckoning what mathematics itself is" (Ibid. 177).

Science as a Theory of the Real

Central in Heidegger's conviction that modern science lacks the possibility to reflect on its own activity is that it must be understood as "the theory of the real" (Ibid. 157). To understand why this definition culminates in the idea that science is unable to reflect, it must be specified which meaning Heidegger attributes to the two central terms in his definition: "theory" and "real."

In German, "real" translates as *das Wirkliche*. It is by specifying the etymological background of this German noun that Heidegger develops his understanding of "real." *Wirkliche* contains the German verb *Wirken* (to work), which has the connotation of action: being real implies to *do* something. According to Heidegger, *Wirken* (working) must be understood as a making present that can be understood both in terms of "something bring[ing] itself forth hither into presencing of itself [and as] the bringing hither and forth of something as accomplished by man" (Ibid. 160). Thus, presencing can rest on either internal principles through which something reveals itself as itself or external principles that depend on human beings doing the presencing in a specific way. It is in the possibility of the latter sense of presencing that the development of modern science must be understood. Because, so Heidegger argues, the presencing of something as accomplished by an external principle has historically developed into the concept of causality: what is real started to be defined as something that was the consequence of a previous cause. Modern science relies on this principle of causality and considers it as its task to precisely determine the relation between the consequence and that which was its cause. This requires having a conception of the real that allows for the humans to intervene in exactly, and to exactly determine the consequences of certain interventions. Accordingly, "the real now shows itself as object, that which [one] stands over against" (Ibid. 162). The specific way of presencing particular to the age of modern science is therefore the re-presenting of something *as* object.

Now Heidegger asks how the re-presenting of the real as an object sphere relates to science as *theory*. Just as he did with the term "real," Heidegger's first step in defining "theory" is etymological. He traces the noun "theory" back to the Greek *theōrein* and holds that this etymology points to an interpretation of theory as the attentive way of looking *by the involved human being*: "It follows that *theōrein* is *thean horan*, to look attentively on the outward appearance wherein what presences becomes visible and, through such sight—seeing—to linger with it'" (Ibid. 163). This meaning of theory disappears in modern science, so Heidegger argues, because it re-presents reality by turning it into a mathematical domain of investigation (i.e., as presented in terms of objectivity). In modern science, theory no longer means looking attentively on what presences itself, but is transformed into an understanding of theory as the observation of the real. It is through this transformation that science today is associated with the disinterested studying of nature. However, "disinterested" observation can only occur *if* nature is represented as a set of objects and only if theory "challenges forth the real specifically through aiming its objectness" (Ibid. 167).

In Heidegger's view, observing nature presupposes a specific setting up of the real, in which nature appears as something that can be passively observed from the outside. In other words, nature must be entrapped as an object, such that a subject-object relation between an observer and nature can come into being. Rather than interpreting observation as an act happening *within* nature that has an openness to the other things in it, a subject with the specific quality of *not* being part of nature ensures the possibility and ongoing existence of modern science. The observing subject must secure that nature is disclosed in this way, because this allows for a delimitation of questions of which it is already known that they can be answered in principle:

> Theory makes secure at any given time a region of the real as its object-area. The area-character of objectness is shown in the fact that it specifically maps out in advance the possibilities for the posing of questions. Every new phenomenon emerging within an area of science is refines to such a points that it fits into the normative objective coherence of the theory. (Ibid. 169)

This does of course not imply that scientists know in advance what the exact answer to their questions will be, but it is presupposed that answers are to be found within the disclosure of nature as mathematical. Every scientific object that appears will appear within this specific conception of nature and will therefore not challenge how science observes nature *as* real.

Reflection as the Limit of Science

But would one not say that theoretical change in science is accompanied with a reconceptualization of nature? Do shifts such as the ones from classical mechanics to quantum mechanics, or from vitalism to Darwinism, not lead to a different constitution of nature? According to Heidegger, these changes within science should not be considered as new constitutions of nature, because these can only occur in the ongoing disclosure of nature as mathematical. For example, he interprets changes from classical mechanics to quantum mechanics not as constituting a new way in which nature becomes present, but as a narrowing of the validity of physics *within* the already existing disclosure of nature. The distinction between the two paradigms in physics he finds "in the experience and determination of the objectness wherein nature sets itself forth. Nevertheless, what does *not* change [. . .] is this: the fact that nature has in advance to set itself in place for the entrapping securing that science, as theory, accomplishes" (Ibid. 172–173). Even though what is taken to be the most fundamental characteristics of what is considered "nature" change, in both physical theories it remains to be conceived as something disclosed mathematically. To deny the calculability of nature would mean to deny the essence of modern science. As such, changes in the set of objects that exist within the mathematical space in which nature is disclosed should not be equated with reflecting on the re-presenting of nature presupposed in modern science.

Similar to his line of reasoning in "The Question Concerning Technology," Heidegger argues that this is the consequence of the fact that in the end, the entrapping of nature as a spatiotemporal coherence of motion is not caused by the activities of individual scientists. The disclosure of nature in terms of its objectness is not to be attributed to science, but "the essence of science is rendered necessary by the presencing of what presences at the moment when presencing sets itself forth into the objectness of the real" (Ibid. 168–69). It is precisely this moment that must be *reflected* upon and points to the need for developing a position that does not a priori equate nature with the way it is mathematical disclosed.

This points to the questions of how the presencing of modern science relates to Heidegger's idea that Technology challenges nature to be revealed as standing-reserve, because

> scientific representation, for its part, can never decide whether nature, through its objectness, does not rather withdraw itself than bring to appearance the hidden fullness of its coming to presence. Science cannot even ask this question, for, as theory, it has already undertaken to deal with the area circumscribed by objectness. (Ibid. 174)

When reflection on science does not occur, nature is unproblematically equated with the objectness in which it is presenced in modern science. However,

> nature, in its objectness for modern physical science, is only *one* way in which what presences [. . .] reveals itself and sets itself in position for the refining characteristic of science (Emphasis mine). [. . .] Scientific representation is never able to encompass the coming to presence of nature; for the objectness of nature is, antecedently, only *one* in which nature exhibits itself. (Ibid. 174)

There is always a surplus in how nature can become present, besides how it is represented through a theory of the real, and because of this, it precisely *is* problematic to unproblematically equate nature with how it is presenced in modern science.

Is Science With or Without Practice?

The dual nature of modern science points to an ambiguity in Heidegger's conception of modern science: on the one hand, he seems to put modern science and the "theory of the real" that grounds it on equal footing with the way

Technology enframes nature to reveal itself as standing-reserve. On the other hand, however, it is only when modern science is considered as a "theory of the real," in which nature "shows itself as object," that can be related to be an observing subject that the accomplishments of science can be accounted for. This leaves an unresolved tension in Heidegger's later work. Contrary to his analysis of the theoretical disclosure of the world in *Being and Time*, science can no longer be interpreted as a practice.

In section 3.1, I have argued that Heidegger's analysis of having a scientific relation with the world rests on two assumptions: (i) for science to function, a mathematical disclosure of nature is required, and (ii) this mathematical disclosure is a mode of disclosure, something that can be obtained in scientific practices. In his later work, the latter of these prerequisites is no longer present; only the mathematical disclosure of nature is presupposed in a scientific worldview. Because this disclosure is no longer grounded in practice, it is no longer possible to interpret it as a rupture with our initial understanding of the world. This is why Heidegger can conclude that only in the modern epoch one can speak of having a worldview in which the only possibility to relate to the world is through its mathematical disclosure. Contrary to his earlier analysis, this is not a special case of world disclosure occurring in a practical context. The challenge left to solve this tension is to understand how the mathematical—*theoretical*—disclosure of the world can be reflected upon in terms of the relations between scientists and the ready-to-hand.

3.4 POSTPHENOMENOLOGY AND SCIENTIFIC PRACTICE

In postphenomenology, the answer to the disappearance of practice in Heidegger's later analysis of science is a return to Heidegger's early phenomenology of tool use. Don Ihde's criticism of Heidegger's monolithic approach toward Technology in "The Question Concerning Technology" is treated as mutatis mutandis also applying to Heidegger's analysis of science. Ihde sees Heidegger's relevance—and the relevance of the philosophy of technology—in the philosophy of science in the inversion of the idea that technology is applied science. Following Heidegger, a key insight of the philosophy of technology, according to Ihde, is the idea that technology is somehow prior to science, that is, to theory. Ihde's starting point is the observation that "scientific knowledge is instrumentally dependent upon technologies. It is 'constructed' through the use of instruments which are *technologies*" (Ihde 1997, 73, my emphasis). The importance of conceptualizing scientific instruments as technologies is not to be surpassed; it allows Ihde to conceptualize scientific instruments as *mediating* the reality that scientists study. "They are the concrete and material operators within scientific praxis" (Ihde 1991, 45), which "*non-neutrally* transform [. . .] the project or object towards which the technology is directed, and reflexively, the human user of that technology" (Ihde 1997, 73–74) (Emphasis in original). Thus, when scientific instruments are understood as mediating technologies, the principle distinction between scientific practices and other practices disappears from view, since both can be equally analyzed in terms of the role that mediating technologies play in them.

Ihde's analysis of the role of technologies in scientific practice has a clear Heideggerian legacy: he holds that technology is ontologically prior to science, thereby turning theorizing into a minor aspect of scientific practice: "Theorizing becomes a special, highly speculative exercise of scientific imagination—important, but both reduced in size and open to greater skepticism—in regions outside the current reaches of instrumental possibility" (Ibid. 100). He argues that "increasingly many scientific phenomena are *technologically carpented* phenomena" (Ibid. 137) (Emphasis in original). This implies that an increasingly large set of scientific phenomena could not be observed when the relevant technologies were not available. Furthermore, he holds that science is likely to follow the inclinations of a technological trajectory, which is related to "the capacities opened up by instruments, capacities of a technological possibility leading to the productive capacities of experimental science" (Ibid. 137). Accordingly, he proposes dropping the notion of science and replacing it with the more appropriate label "technoscience." Thus, Ihde's analysis of science can be understood in terms of the *Primacy of Scientific Practice* claim, because he holds that scientific knowledge is bound up with certain practices and that these practices are thoroughly technological and thus cannot be intelligibly discussed science without reference to these practices.

Mediated Perception: Scientific Instruments as Disclosing Reality

Ihde gives a substantial twist to the late Heidegger's understanding of the ontological priority of technology, when understanding the priority of technology in science in terms of how reality is made present by concrete *technologies* (i.e., scientific instruments). Ihde dismisses Heidegger's later analysis of science by considering it on par with his analysis of technology and conceives of it as an unfounded romanticism that science cannot but reveal the world as a standing-reserve (1991, 58; 2010, 74–85). While he takes up the late Heidegger's idea that technologies reveal the world in a specific way, he does not do so by taking "Technology" as an epochal way of disclosure, but is instead interested in how particular technologies reveal the world in specific ways in scientific practice. By moving back to Heidegger's phenomenology of tool use, he is interested in how the world is constituted in relation with the ready-to-hand. From this perspective, technologies open a domain in which a set of scientific objects appear that can be intentionally related to. In Ihde's case, it is technologies that allow to disclose the world in new ways. Analyses of scientific practices should accordingly focus on, for example, how the telescope shaped the development of astronomy, or how our understanding of the human being is shaped through the use of brain imaging technologies (e.g., Ihde 2009, 51–61).

Given that postphenomenology focuses on human-technology *relations*, an analysis of technological practices does not end with an analysis of the technologies involved, but must also consider the human subjects relating to them. After all, it is concrete individuals such as Galileo Galilei who point a telescope to the moon and collectives of cognitive neuroscientists who use brain imaging technologies to investigate human cognition. To articulate how scientific instruments shape the world that scientists observe, Ihde appeals to Maurice Merleau-Ponty's phenomenology and departs from the noncontroversial assumption that a fundamental condition of scientific observations is to bring the phenomena to be observed within the domain of human perception, and with Merleau-Ponty, he stresses that perception always implies *embodied perception*. However, secondly, he also draws on Merleau-Ponty to argue that interpretations of the world are transformed by *how* the world is perceived. Consequently, how the world that becomes both present and a candidate for interpretation is dependent on the specific technologies through which human beings relate to the world. Merleau-Ponty illustrates this with an example of how the world appears to a blind man using a cane to orient himself to his environment. When doing so, the point of the stick

> has become an area of sensitivity, extending the scope and active radius of touch, and providing a parallel to sight. [. . .] The position of things is immediately given through the extent of the reach which carries [the blind man] to it, which comprises besides the arm's own reach the stick's range of action. [. . .] [This] is knowledge in the hands, which is forthcoming only when bodily effort is made, and cannot be formulated in detachment from that effort. (Merleau-Ponty 2012, 144)

The case in point is here that when the blind man uses his cane, there is only a world present to the extent that it becomes present through the stick, thereby making any experience and interpretation of this world contingent on the stick.

According to Ihde, this idea can be extrapolated to how the world becomes present to scientists when they relate to it through specific technologies, because "the cane fall[s] into the same existential use as many scientific instruments" (Ihde 1991, 30). This is why Ihde holds that science is not "abstract" and "theoretical" but should rather be considered in terms of how the reality investigated in scientific practices comes into being through the scientific instruments that are used. This requires focusing on *how* scientific instruments bring the phenomena of science within the domain of embodied perception.

Ihde characterizes science as a fundamentally visual practice and therefore understands it in terms of a *visual hermeneutics* (Ihde 1998). The primary way in which scientific phenomena are made visible to scientists is by turning them into readable images. In Ihde's view, the images that scientists are confronted with allow for multiple coherent readings; they embody a *multistability*. In other words, the images that scientists are confronted with are not immediate depictions of reality but can be interpreted in multiple coherent ways. As a consequence, a scientist must engage in a hermeneutic act: images "must be 'read,' 'decoded,' to detect [. . .] what is being presented" (Ibid. 67). Analogous to an interpretation of a Necker cube, there are varieties of interpretations of a scientific image possible that exclude each other but are in principle equally valid (Ihde 2009, 56). Accordingly, a continuity between ordinary practices and scientific practices can be found in how images are brought within the reach of visual perception, thereby instantiating multistability. Accordingly, Ihde makes instrumentally mediated *perception* the central category in his analyses of scientific practices.

This brief analysis of Ihde's understanding of the role of technologies in scientific practice allows the articulation of in what sense he subscribes to the *Primacy of Scientific Practice* claim: precisely because he defines scientific practices as technological practices, science must be understood in terms of the relations between scientists and technologies that appear as ready-to-hand, thereby mediating the reality that scientist study. Furthermore, he characterizes the relations between scientists and technologies as ones in which technologies mediate human *perception*, thereby interpreting the priority of the ready-to-hand in terms of how they transform the phenomena to be observed into phenomena that are susceptible to an embodied subject. Accordingly, the priority of the ready-to-hand must be understood in terms of the priority of embodied perception in scientific practice.

3.5 RELATING TO THE PRESENT-AT-HAND: SCIENCE AS A SPECIFIC KIND OF PRACTICE

Ihde thus transforms the late Heidegger's analysis of Technology into a continuation of his phenomenology of tool use. However, the former he considers a "romantic amplification of the deeper inside from the [ready-to-hand] period" (Ihde 1998, 55), in which the ready-to-hand is attributed a metaphysical meaning in which it—as generalized—threatens to reveal nature as standing-reserve. Despite this later—in Ihde's view—unmotivated romanticism, he urges to not dismiss Heidegger's work completely, because his later work "does not mitigate the early insights involving technology as a way of *seeing*" (Ihde 1991, 52, my emphasis). Accordingly, Ihde translates what Heidegger calls a technological "way of revealing" into a technological "way of seeing," through which the priority of technology over science is affirmed. But to what extent does this translation allow an understanding of how scientists establish a relation with the present-at-hand (i.e., to also consider the distinct *scientific* character of scientific practice)?

The primary advantage of the postphenomenological criticism on the later Heidegger is that it opened up the possibility to understand the impact of technologies on our lives through our interaction with them on a concrete level. When interpreting technologies as being ready-to-hand to human beings, it becomes possible to examine them as tools through which a world is revealed to in the relationship with them. The transparency of technologies in our ready-to-hand relation with them allows for establishing both an intentional relation with a phenomenon and the development of a specific hermeneutic strategy to interpret it.

Analogous to how one might relate to a nail *through* a hammer, one might relate to a brain *through* a brain scan (e.g., Ihde 2009, 43). This paves the way for the postphenomenological idea that the Heideggerian analysis of either Science or Technology as enframing nature should be dismissed. Just as in ordinary practices, technologies in science allow relating to various new scientific objects. It is exactly this openness of concrete technologies that can be seen as instantiating the continuum between ordinary and scientific practices. Ihde takes this to be the central meaning of Heidegger's interpretation of Technology as a way of revealing: technologies in scientific practice establish relations with different scientific objects in myriad ways, because they literally allow to *see* the world differently. This is an extension of how Heidegger discussed a primary characteristic of tools being that they withdraw when put to use. Similarly, in mediating the relationship with the world, the relevant technology dissolves in how it allows to relate to a certain scientific object. Only when a technology stops functioning within this context does it appear as an object. In other words, Ihde limits Heidegger's analysis of the phenomenology of tool use to the *breakdown mode* of establishing a relation with the present-at-hand.

Technologies as Mediating the Deliberative Mode of Relating to the Present-at-Hand

In scientific practice, it is not technologies but something else that is disclosed as *present-at-hand*, namely, the scientific object mediated by the technology that is transparent. Indeed, as Ihde argues, precisely because of the transparency of scientific technologies, scientists can focus on interpreting images rather than being explicitly engaged with the process of their creation. Hence, he identifies the reading of such images as subject to a similar mechanism as how one relates to a Necker cube (i.e., such images can be interpreted in multiple coherent ways). Accordingly, so Ihde argues, technologies such as fMRI mediate how a scientist can relate to the brain. However, when examining concrete scientific practices involving brain scans, it turns out that it is scientific objects such as diseases or psychological traits that are related to *through* a brain scan. For example, a medical doctor might use MRI to detect a brain tumor, or a cognitive neuroscientist might use fMRI to study a scientific object, such as visual attention. In these cases, strictly speaking, it is not "the brain" that is related to through a scientific instrument but some scientific object that is transformed when related to through a brain scan. These transformations are not immediately perceptually given but require an existing conceptual apparatus that can be found in theories within the relevant scientific domain.

Because of the multistability involved when reading images, it has been stressed that scientists must apply *hermeneutic strategies* to make sense of what is being presented and that (i) these images allow for multiple hermeneutic strategies and (ii) these strategies may yield conflicting interpretations (Ihde 2009; Rosenberger 2011). This idea serves as a way to understand conflicts between scientists as grounded in having developed different hermeneutic strategies to interpret a technologically mediated phenomenon. Accordingly, by continuing to focus on how technologies mediate human perception, one cannot answer the questions of *why* a specific hermeneutic strategy is thought to be appropriate or *why* this specific strategy leads to specific conclusions. A focus on the technological mediation of *perception* neglects that a variety of scientific objects can be made present through a brain scan and that theoretical knowledge is required to be able to perceptually differentiate between these different objects. This points to the need to not only understand the role of scientific instruments in terms of the mediation of perception but also focus on how the world is disclosed in *scientific* practice, as mediated by technologies.

As already emphasized in this chapter, this practice is precisely regulated such that the world can be disclosed as a domain of objects that can be measured. I have termed this practice the *deliberate rupture mode* of relating to the present-at-hand through the mathematical disclosure of the world. Within this mode, the technical setup of an experiment and the preparations of microscopic observations can be interpreted as practical concerns both necessary for a specific form of world disclosure and as mediating new forms of world disclosure. The mode of disclosing the world as present-at-hand through a deliberate rupture differs from Ihde's interpretation of Heidegger's phenomenology of tool use. Firstly, it emphasizes that technologies in scientific practice function to mediate a specific *mathematical* disclosure of the world. Secondly, scientific instruments serve not only to let all entities dissolve in the whole of the ready-to-hand but also to make objects specifically appear as present-at-hand.

What Heidegger's hermeneutic analysis of science indicates is that the mediating role of scientific instruments must be understood not only as instantiating a specific way of seeing but rather as mediating how the world in scientific practice is revealed by co-constituting the observing subject to which the world is revealed. This is stronger than Ihde's idea that scientific instruments give rise to new ways of perceiving reality in the world. Rather, it takes technologies not only to reshape how the world is seen but also to change the domain of objects that this world consists of. Consequently, "world" in this context refers to the mathematical disclosure of the world in scientific practice, which implies that the domain of objects within this world do only relate to each other within how the world is disclosed in scientific practice.[4] Precisely because technologies in scientific practice function within the *deliberate rupture mode*, they are constituted in terms of how the world is disclosed when it is explicitly attempted to establish relations with present-at-hand object. In analogy with Heidegger, one can say that technologies primarily *reveal* new domains of objects rather than instantiating new ways of seeing in the first place.

3.6 CONCLUSION: TECHNOLOGICAL MEDIATIONS AND THE THEORETICAL DISCLOSURE OF NATURE

If scientific practice is to be interpreted as presupposing a mathematical disclosure of the world and as a practice in which technologies reveal new domains of objects, there is both a crucial similarity and a crucial difference with Heidegger's understanding of Technology as a way of revealing. On the one hand, this idea departs from the universalist conception of Technology as Enframing nature, as technologies are interpreted as embedded in scientific practice and attain their meaning within the practice that I have interpreted as aiming to constitute a deliberate rupture with our initial contact with the world. On the other hand, however, the mathematical disclosure of nature in scientific practice comes close to the idea that for science to function, the Enframing of nature must be presupposed. This seems to imply that in this model it is Science, rather than Technology, that holds the danger of challenging nature to reveal itself as standing-reserve.

When returning to Heidegger's analyses of Science in "The Age of the World Picture" and "Science and Reflection," a similar dilemma is encountered because both leave open the question whether or not there is a way to interpret Science as a manner of disclosing the world mathematically without reducing nature to this specific manner of disclosing. This is where the postphenomenological criticism of Heidegger's conception proves useful. In stressing that neither Technology nor Science are forces exerting influence on society from the outside, postphenomenologists argue that a focus on concrete technologies and how scientific objects are revealed through them only attains meaning within practice.[5] Interpreted in this way, the notion of Science must be replaced by "scientific activity" or "scientific practice," notions referring to both the practical character of science and the specific nature of this practice, namely, the theoretical disclosure of the world.

When connecting this with Heidegger's idea that Science or Technology must be interpreted as the enframing of nature, or of the setting-upon of a scientific worldview onto nature, it seems that the mathematical disclosure of nature is indeed "a normative theory of nature" (Heidegger 1977b, 169). However, instead of taking this as a negative consequence of modern science per se, it is also possible to interpret this normativity in terms of the *deliberate rupture mode* as developed in *Being and Time*. From this perspective, scientific practice does indeed depend on a normative theory of nature, and it is precisely this theory that creates the rupture with ordinary practices. Contrary to ordinary practices, where entities only become present-at-hand because of their malfunctioning, scientific practices are structured to disclose nature *as* a set of entities.

Therefore, when investigating the role of technologies in scientific practice, it must be acknowledged both that it is a general feature of scientific practice to attempt to constitute a rupture with ordinary practice insofar that it discloses nature as a set of present-at-hand entities, and that this by no means eliminates the practical character of this act. What I take to be the primary contribution of Ihde's reading of Heidegger is his idea that technologies embody a specific way of revealing, such that scientists establish relations with objects only through the use of a specific technology. I have extended this idea by arguing that these technologies do not only instantiate new ways of seeing but open up new domains of objects that can be related to, both theoretical-conceptual and practical-perceptual. As such, the set of entities that nature in its mathematical disclosure consists of is by no means fixed but is in constant flux because of changes in the technologies we use to disclose it.

"To the things themselves!" is Husserl's often quoted imperative. But what are the things in scientific practice? Answering this question requires investigating what the objects are that are intentionally related to in scientific practices and how these objects come into being. Accordingly, "the things" toward which I turn in the coming chapters have two distinct meanings. On the one hand, "things" are interpreted in terms of Heidegger's *deliberate rupture mode* (the scientific objects that scientists relate to). On the other hand, the concrete technologies in scientific practice are interpreted as "things" that open up new scientific phenomena. In doing this, I hope, by building on both Heidegger's idea that technologies can be understood as a way of revealing within a mode of deliberate rupture and Ihde's postphenomenology of instrumentally mediated perception, to clarify what relation between "science" and "practice" constitutes a "scientific practice."

NOTES

1. In this regard, I align with Glazebrook (2000b), who argues that although the possibility of the scientific disclosure of the world has been central to Heidegger's thinking from the start, but that how this relation is interpreted differs. She points out that, contrary to his analysis in *Being and Time*, the later Heidegger does no longer understand science as constituting one possible relation with nature, as it becomes part of the necessary technological disclosure of the world per se. For a detailed discussion, see Glazebrook (2000a, 5–9) and Wolf (2003; 2005, 139–71, 231–40).

2. In the English word "research," the Heidegger idea becomes immediately clear. The "re-" implies that something must be found in a pre-given space and is stipulated as already there.

3. Unless indicated otherwise, "mathematics" should in this chapter be understood in terms of Heidegger's understanding of the Greek *Ta mathêmata*.

4. Note that this does not necessarily imply that they do not correspond to objects in our daily practice, but neither suffices as an explanation why they may or may not do so.

5. Two consequences can be drawn from this move: First, it allows us to ask for the role of technologies in scientific practices. Secondly, it allows for appreciating the role that scientific discoveries play in the structuring of daily practices. In this book, I am primarily concerned with the first of these implications.

Chapter 4

To the *Scientific* Objects Themselves

Gaston Bachelard's Phenomenotechnique

INTRODUCTION: THE SCIENCES AND THEIR OBJECTS

In the previous chapters, I have argued that to answer how technologies function in scientific practice, both the scientific and the practical character of scientific practice must be considered. In Being and Time, Heidegger emphasized that the distinct scientific (theoretical-mathematical) character of science is grounded in a specific practice aiming for a deliberate rupture of our everyday relation to the world. In his later works, however, this practical ground of science is absent, making him think of science as terms of an enframing. This would imply that science, just as technology, must disclose nature as standing-reserve. Furthermore, my analysis of the role of scientific instruments in scientific practice in postphenomenology shows that while this branch of thinking does consider scientific instruments as functioning within practices, it does not consider the *scientific* character of scientific practice. In other words, while acknowledging the importance of technologies in every aspect of science, a postphenomenological approach neglects the importance of how this practice is structured around a theoretical disclosure of the world. In this chapter, I begin answering the question concerning the role of scientific instruments in science by considering both of these aspects of scientific practice.

Doing so forces to ask to what kinds of objects scientists relate to and how they differ from the objects encountered in other practices. Intuitively, one thinks of an object that is something concretely tangible (a chair, a table). In the context of science, people tend to share a similar intuition: brain scans allow us to understand *the* brain in a new way, and telescopes allowed Galilei to see *the* moon in a new light. However, the objects of scientific investigation rarely resemble the objects encountered in everyday activity. In the case of the neurosciences, it is often not *the* brain that is investigated, but rather the mathematical relation between a certain behavioral task and the electrical activity in a certain brain area (in the case of EEG). As such, the intentional object to which a scientist relates within this investigation is not the brain as such but rather the relation between a behavioral task and electrical activity, which can be visualized in a brain scan.

As I have attempted to clarify in the previous chapter, Don Ihde largely dismisses the importance of theory in contemporary scientific practice and emphasizes the importance of observational technologies in the construction of scientific objects. This, so I argued, leads him to neglect the Heideggerian insight that a deliberate rupture with the ready-to-hand must take place for a relationship with the present-at-hand to be established. This is reflected in the importance in his philosophy of Merleau-Ponty, who emphasizes the importance of perception in the experience of the human subject. Following Michel Foucault, David Hyder, and Hans-Jörg Rheinberger (Foucault 1989; Hyder 2003; Rheinberger 2005), I consider Merleau-Ponty as part of one line of phenomenologically inspired thinking about science in France that stressed the subjectivist side of scientific perception. The other line in this tradition is more concerned with how scientific objects are grasped *conceptually* and can be associated with philosophers as Gaston Bachelard, Jean Cavaillès, Georges Canguilhem, and Michel Foucault.

The present chapter is concerned with one of the earliest representatives of this latter tradition, Gaston Bachelard, who was one of the first to note both how technologies shape the reality studied in scientific practice and how they help providing conceptual access to this reality. A central feature of Bachelard's epistemology that is of interest for the purposes of this chapter is his idea that scientific objects are not "real" in an everyday sense. Rather, scientific objects emerge within a *phenomenotechnique*: they come into being as the consequence of their relationship with the available scientific instruments, existing scientific theories, and the experimental setup in which they appear. Consequently, the things to which scientists relate form a special class of objects that manifest within the phenomenotechnique.

In this chapter, I attempt to show how this concept is grounded in a break between scientific and ordinary experience and how Bachelard attempts to connect scientific practice with scientific theory through it. I do so by introducing Bachelard's idea of the epistemological rupture between everyday experience and scientific experience, and analyze how this break is realized in scientific practice through a phenomenotechnique. Then, I critically scrutinize the meaning of the "-technique" in "phenomenotechnique" and show that the concept does not primarily refer to the technological creation of scientific phenomena in practice but rather concerns the realization of an objectively rational environment in which science must take place. As a consequence, so I argue, Bachelard's epistemology is primarily of help when we wish to understand what is specifically scientific about scientific practice but seems less useful when referring to the practical aspects of scientific practice. I propose a postphenomenological reconceptualization of phenomenotechnique as a *phenomenotechnology* and argue that this allows a constitution of the rationality of science *in practice* to become a topic of (post)phenomenological investigation.

4.1 THE EPISTEMOLOGICAL RUPTURE: SCIENCE AND EVERYDAY LIFE

A central concept of Bachelard's philosophy of science is the "epistemological rupture," which poses a fundamental divide between science and everyday life, and hence between scientific experience and ordinary experience. Bachelard holds that science should overcome the "obstacles that everyday life has set up" (Bachelard 2002, 28). He juxtaposes the imaginative character of everyday life with the rationality of science and argues that science should break with our natural experience to function properly.

Bachelard's philosophical work is structured around two primary themes: science and poetics. On the one hand, he aims to show how and why science must get rid of everyday experience to be able to engage in rational activity. On the other hand, he attempts to develop an account of how everyday life experiences can give rise to meaningful experiences and how these experiences are translated into poetic images. In the context of science, however, everyday experiences appear as obstacles that must be overcome. But why does the distinction between science and poetry lead to a distinction between everyday experience and scientific experience, and why does everyday experience have no place in science?

The idea that science must break with ordinary experience is grounded in two assumptions: on the one hand, Bachelard holds that the objects of science do not coincide with the objects encountered in everyday life (Chimisso 2015, 65). On the other hand, he argues that primary experiences are always primitive in the sense that they refer back to subjective forms of reverie and therefore have no place in rational activity. Of course, human beings cannot do away with their subjectivity entirely: "Even the scientist, when not practicing his specialty, returns to the primitive scale of values" (Bachelard 2002, 4). However, when scientists do practice their specialty, the danger of falling back into primitive experience remains present, because the scientific mind "takes up the same primitive themes time and again and always operates as it would in primitive minds, and this in spite of the successes of systematic thought and even in face of the findings of scientific experiments" (Ibid. 4). But through what kind of primitive drives is reality ordinarily experienced, and why are they hindering the development of science?

Bachelard's Poetics

Before specifically addressing the question posed above, it is necessary to take a short detour through Bachelard's poetics to clarify why he holds that everyday experience is an obstacle for scientific thinking.[1] Contrary to the rationalist purpose of science, the initial contact with the worlds is always one of reverie and dream. The primitive encounter with an object induces a reverie that transforms initial experience into an image that transcends the object that it is concretely related to. This transcending occurs through our primitive values and precedes any kind of reflection: "We are being faithful to a primitive human feeling, to an elemental organic reality, a fundamental oneiric temperament" (Bachelard 1999, 5). Only when overcoming this oneiric state can a rationalist stance toward the world come into being: "In point of fact I see no solid basis for a natural, direct, elemental rationality. [. . .] Rationalist? That is what we are trying to *become*" (Ibid. 7) (Emphasis in original).

This already indicates that neither can primary experience of the world be taken as the foundation for rationality nor can rationality be founded in the transformation of primary experiences. For Bachelard, such transformations can only be described, leading to poetic images. These images do not say something about reality itself; rather, a poetic image is poetic to the extent that it is supra-natural. Hence, imagination "is not the faculty for forming images of reality; it is the faculty for forming images which go beyond reality, which *sing* reality" (Ibid. 16) (Emphasis in original). Thus, for Bachelard, poetry is the forming of a linguistic manifestation of the images induced by the reveries that one engages in when relating to objects in primary experience.

The project intending to describe the relation between poetry and the initial relation with the world is sharply distinguished from Bachelard's project of overcoming the primitive relation with the world in his epistemological works. For example, in *The Flame of a Candle* (2012), he writes that this essay on our primary relation with humanized fire should not be confused with his epistemological project aiming for the establishing of objectivity in science: "The history of combustion remains, until Lavoisier, a history of pre-scientific views. The examination of such doctrines belongs to a psychoanalysis of objective knowledge. This psychoanalysis must get rid of images to determine how ideas are organized" (Bachelard 2012, 8).[2] The primary difference between these projects is precisely that Bachelard's poetics is interested in how images established through reverie can attain meaning in poetry, while this notion of image is precisely something to be avoided when scientific knowledge is considered. In summary, Bachelard's poetics ask for how poetic imagination *derives from* ordinary experience, while his epistemology aims to show how the rationality of science must be obtained through a *break with* this ordinary experience.

Everyday Experience as an Epistemological Obstacle

To make clear why he maintains a clear distinction between everyday experience and scientific practice, Bachelard introduces the notion of "epistemological obstacle." The epistemological obstacle of ordinary experience is grounded in how human beings experience reality *psychologically*. This indicates that Bachelard's epistemology is in effect a normative theory of rationality (Bachelard 2002, 27). The normativity in Bachelard's epistemology takes science that is presently accepted as the standard of rationality, and it is from this perspective that specific obstacles are identified. Bachelard's discussion of these obstacles has a twofold aim: on the one hand, he aims to show that the rationality of the scientific mind differs from the prescientific mind because it has overcome existing obstacles in its own development. On the other hand, he holds that each individual scientist must overcome the epistemological obstacles present in her own psychology (Ibid. 20). His general principle for establishing rationality in science is that it "must be formed *against* nature, against all that comes from nature's impetus and instruction, within us and outside us, against natural allurements and colourful, diverse facts" (Ibid. 33) (Emphasis in original). When remaining too much attached to nature, specific obstacles are encountered that should be avoided: *verbal* obstacles, *substantialist* obstacles, and *animist* obstacles. The prescientific mind is too attached to using ordinary language in science (verbal) to reduce the different qualities of a phenomenon to a single substance (substantialist) and to interpreting the workings of scientific phenomena in terms of biological processes in the human body (animist) (Bachelard 2002).

Let me explain how Bachelard conceives of these obstacles by showing how he invokes the substantialist obstacle to argue against the validity of Descartes's wax thought experiment. Descartes argues that the wax cannot be clear and distinctly be perceived by the senses when its form has changed when it is heated, "for whatever came under the senses of taste, smell, sight, touch or hearing has now changed; and yet the wax remains" (Descartes 1993, 21). This is why he concludes that the continuous nature of the wax, namely its *extension*, is not grasped by the senses but by the mind alone. Bachelard, however, argues on the contrary that the diverse manifestations of the wax under different circumstances are the basis of every form of scientific research, because they show the complexity of the wax. In other words, it is only through the complexity of its objects that science can be formed against our natural inclination to think of wax as an object in our ordinary experience.

Thus, while Descartes concluded that the only certainty concerning the wax should be found in *simplicity* (extension), Bachelard holds that scientific certainty should be found in the *complexity* of the shape of the wax under different circumstances.

More concretely, Bachelard's criticism amounts to Descartes's mediations being grounded in the individual subjectivity of the philosopher and not coinciding with the scientific aim of objectification. In Bachelard's view, the phenomena of science can never exist in the mind alone but has a twofold relation with the empirical. Firstly, the world of science is made complex in scientific experiments that force us to acknowledge the complexity of scientific phenomena. Secondly, it is only in relation to this increasing complexity that science can overcome the subjective and imaginary aspects of ordinary experience. When comparing Descartes's way of approaching the wax with contemporary laboratory science, he argues that "subjective [Cartesian] meditation is bent on attaining clear and definitive knowledge; objective meditation differs from this by the very fact that it makes progress, by its intrinsic need always to go further, to extend the limits of the known" (Bachelard 1984, 171). This objective meditation of the laboratory scientist marks an epistemological rupture with how nature is encountered ordinarily.

It is at this point that it becomes clear why Bachelard holds that Descartes's meditations fall prey to the substantialist obstacle. While Descartes concludes that the changing qualities of the wax must lead to a dismissal of the trustworthiness of sensory experience, for Bachelard, it is, on the contrary, precisely this experimental diversification that grounds the objectivity of the sciences. Because Descartes continues to interpret wax as a substance, he fails, so Bachelard argues, to note that certainty about wax can only come into being because of its diversification. The scientific mind must be open to the new experience of a new reality and rationalize the complexity of scientific phenomena by relating them to each other. The objective meditation strives to continuously expand the limits of the known through the rational incorporation of experimental diversity.

4.2 BACHELARD'S PHENOMENOTECHNIQUE

As hinted at in Bachelard's discussion of Descartes's wax experiment, it is only in comparison with contemporary scientific practices that earlier types of inquiry can appear as irrational. Similarly, the specific epistemological obstacles necessary to overcome when practicing science are themselves defined by how science is presently practiced. Hence, Bachelard does not refer to any standards of rationality that are external to scientific practice and transcend the limits of time and space. Only by examining up-to-date scientific practices themselves can philosophers sense the rationality that scientists must strive for.

The beginning of the twentieth century saw many great developments in physics. It saw the birth of relativity theory and quantum mechanics and broke with earlier theories in physics that were grounded in classical mechanics. Science was no longer an activity of describing and understanding the objects as they appear to us. Rather, "its purpose is to amplify what is revealed beyond appearance" (Bachelard 2002, 13). This revealing can only occur in specific experimental conditions that cannot be cut loose from the scientific instruments used to create them. Scientific objects are never mere phenomena but are always the product of observed phenomena to which one can only relate to through the instruments, concepts, and scientific theories—it is through these that science can amplify what is revealed *beyond* what is empirically observable. In Bachelard's words, scientific objects are constituted within a *phenomenotechnique*.

Philosophy from Science

As already noted, in Bachelard's view, any philosophy of science must take into account the most up-to-date scientific discoveries and practices. The development of relativity theory and quantum physics opened a new era in science, which should be understood as a clear break with all earlier manifestations of scientific activity. According to Bachelard, this break did not just have scientific consequences; it also forced philosophy to "modify its language if it is to reflect the subtlety and movement of contemporary thought [. . .], [because] science in effect creates philosophy" (Bachelard 2002, 3). Consequently, *if* the findings of the most recent scientific developments contradict traditional epistemological models, these models should be revised. According to Bachelard, the developments within physics indeed contradicted existing philosophical models, as they (i) made the strict distinction between realism and idealism obsolete and (ii) made it impossible to strictly distinguish between a subject that observes an object and an object that is observed by a subject.

The most prominent site of analysis for observing the differences between twentieth-century physics and previous theories in physics is microphysics in the case of Bachelard. According to Bachelard, philosophers searching for the foundations of certain knowledge were either realistic (knowledge is the product of the relation between the subject and the world) or idealistic (knowledge is the product of the human mind) terms. Bachelard aimed to construct a middle-ground position between these positions on the basis of his analysis of the scientific practice of microphysics, which "makes rational entities real" (Ibid. 13). It is in this context that he develops his notion of phenomenotechnique to show that scientific objects are dependent on the rationality of the scientific mind but only become accessible in their empirical manifestation.

Scientific phenomena, such as the Zeeman effect, but also entities such as electrons, so Bachelard holds, cannot be understood in terms of our ordinary experience of nature, as their manifestation depends on a structure that is fundamentally different from our daily encounter with reality. According to Bachelard, such developments within physics and chemistry at the start of the twentieth century mark a break with earlier physical theories. He argues that we should interpret contemporary physics and chemistry as ways of thinking that instantiate an epistemological break with our ordinary way of knowing (Bachelard 1949a, 102). In pre-twentieth-century physics, the scientist was still "one of us" (Bachelard 2005b, 73), in the sense that scientists spoke about the objects present in our ordinary reality. However, as physics was no longer in accordance with our ordinary experiences, Bachelard held that both of these positions were no longer tenable and should be replaced with an epistemology grounded in the actual practice of science and reflect the non-ordinary domain theories in physics are about.

The Phenomenotechnical Realization of Scientific Objects

A key insight in Bachelard's epistemology is that the objects spoken about in scientific practice cannot be cut loose from the instruments used to observe them. Accordingly, one thing his epistemology must account for is that scientific objects are created by the technologies used to observe them: "We must demonstrate that what humans create through scientific technologies does not exist in nature, and is not the natural cause of a natural phenomenon" (Bachelard 1949a, 102). Consequently, the observation of scientific phenomena necessarily coincides with their creation; only when understanding the object's process of becoming can a scientist understand what he has created and subsequently observed. The objects of science are in constant development, as they are dependent on how they are technologically constituted, and also the objects of the different sciences are different, as each of them employs a particular set of technologies, with their own set of accompanying objects. Hence, he advocates a pluralistic mindset within epistemology that is inspired to consider "each hypothesis, each problem, each experiment, [and] each equation" (Bachelard 1968, 12). Reflecting on these concrete instances occurring in scientific practice is the foundation of Bachelard's new epistemology that breaks with the traditional dichotomies of empiricism/idealism and positivism/rationalism.

A central concept in this new epistemological framework is the idea of *phenomenotechnique*, the idea that we cannot understand our experience of scientific objects without an understanding of how they are technically realized (Bachelard 1949a). According to Bachelard, what is unique about the objects of twentieth-century physics is not only that they must be realized but that their coming into being cannot be understood without reference to the realization technique that accompanies them. On the one hand, this refers to the scientific instruments that are needed to produce phenomena while, on the other hand, it concerns how these phenomena are dependent on theoretical developments. This is why he speaks about the objects of science in terms of a phenomenotechnique:

> It is then that we understand that science *realizes* its objects without ever just finding them ready-made (Emphasis in original). Phenomenotechnique *extends* phenomenology. A concept becomes scientific in so far as it becomes a technique, in so far as it is accompanied by a technique that realizes. We are therefore very conscious that the problem of modern scientific thought is once again a problem that is philosophically intermediate. As in Abelard's time, we would like to take up an intermediate position between realists and nominalists, positivists and formalists, between advocates of facts and of signs. (Bachelard 2002, 70)

This notion emphasizes how scientists always relate to the object under study through a technique by which it is constructed. Science cannot be cut loose from the phenomenotechnique in two ways, as it (i) embodies a technological mode of action in the sense that it produces phenomena, and (ii) these phenomena only become manifest through the materiality of scientific instruments.

It is important to note that for Bachelard, this concept is not limited to material technologies. Most importantly, he considers mathematics a technique capable of realizing new scientific objects. This is the second crucial difference instantiated by twentieth-century physics: scientific objects are no longer in continuity with ordinary experience, as they are mathematical relations rather than concrete substances. In other words, scientific objects can only be considered real insofar they are rationally—mathematically—related to, and the only way to achieve rational objectivity "is to set forth, in a discursive and detailed manner, a method of objectification" (Ibid. 12). Hence, the phenomenotechnique also requires the present of theoretical commitments, because it is through these that a particular mode of action is taken and through which the working of material technologies can be manipulated. Hence, the objects of science are both technologically and theoretically invested, and to understand how they are realized is to understand the interplay between technology and theory.

The fact that scientific objects are made present within a phenomenotechnique must give rise to a new philosophical position in between rationalism and empiricism: "In relativity, the terms of the dialectic are rendered solid and cohesive to the point of presenting a philosophical synthesis of mathematical rationalism and 'technological' empiricism" (Bachelard 1949b, 565–66). At other places, he refers to this middle position as an "applied rationalism" or a "technical materialism." These terms illustrate that a philosophy of science should always oscillate between the rationality of science itself that creates a domain of inquiry that breaks with ordinary experience and the objects that are present in this domain that are the product of technological manipulations.[3]

Technologies within the Phenomenotechnique

For Bachelard, the importance of technologies becomes most apparent when considering that, in twentieth-century physics, it is impossible to strictly distinguish between scientific phenomena and the experimental setups in which they appear. He connects this issue to the increasing need for precision in scientific practice, which can only be obtained through the use of technologies through which precision can be guaranteed (Bachelard 2005a). In addition, he holds that this increasing need for precision realized through material technologies is of crucial influence on how scientific thinking develops as the testing and development of hypotheses needs to be coordinated from the available instruments. Whereas Bachelard argues that theories and the testing of them cannot be meaningfully separated, he does not consider scientific technologies as the primary creator of the reality that scientists study. Technologies function *within* the phenomenotechnique and accordingly are closely connected to theoretical developments.

He illustrates this point with reference to both scientific experiments and measuring instruments. Measuring instruments such as thermometers can change the line of questioning regarding the phenomenon of heat, as they stabilize this phenomenon and allow for its increasingly precise standardization. They are both the product of the existing theoretical knowledge concerning heat, as a manifestation of the technological opening-up of a new line of physical research; as an object, the thermometer provides an increasingly objective manifestation of the phenomenon of heat.[4] Similarly, the phenomena observed in experiments, such as electrons, cannot be detached from how they are humanly observed, indicating the dependence of scientific theories on our technological means to test them. This is why the concept of phenomenotechnique must be introduced; it is only in the specific techno-epistemic locus of a specific science that scientific objects attain their meaning.

While it is tempting to interpret the standardization of heat and our capability to experimentally detect electrodes in terms of science as cumulative progress toward scientific truth, Bachelard's view of scientific progress does not validate such a conclusion. Although he believes that state-of-the-art science should be understood as the standard for rationality, this rationality is not guaranteed with reference to its capability of discovering the truth about something that is external to it. The standardization of heat through the thermometer extends the range of phenomena that science must theorize about, as "temperature is seen on a thermometer, one does not feel it. Without theory, one would never know whether what is seen and what one feels correspond to the same phenomenon" (Bachelard 1968, 9). Hence, a change in the technical structure of the phenomenotechnique forces scientists to develop scientific theories to account for the existence of two different ways of phenomenologically accessing the same object and must show that both of these ways establish a relation with the same object—and this requires a rational, theoretical foundation.

Thus, even though Bachelard argues that "when we have followed contemporary physics, we have departed nature to enter in a factory of phenomena" (Bachelard 1951, 10), one should not conceive of this analogy as diminishing the distinction between scientific practices and the practices of mass production in industrial factories. Scientific practices are dependent on the uniform working of scientific technologies, which are interpreted as "reified theorems" (e.g., Bachelard 1949a, 103) that "provide a certitude which is lacking in more passive knowledge" (Bachelard 2005a, 180). Technologies provide the basis for the construction of phenomena that remain stable in different experimental contexts in a similar way that mass production creates a stability among a set of products. However, while the maintaining of this stability is a sufficient condition for pursuing the ends of technological thinking, in scientific practice, this stability must be transcended to gather new knowledge.

The transcending of the stability of the given phenomenon is what Bachelard refers to as an "unanticipated paradox which shows knowledge to be fleeting and mobile, and the action which it brings to light to be solid and assured" (Ibid. 176). According to Bachelard, the crucial difference between science and industry must be located in the end of a technological process for industrial means being part of how it is realized, while this goal directedness is absent in science. Industry deals with closed systems that can be explained in terms of their functional working. For example, the working of a waterwheel that produces electricity can be exhaustively explained in terms of the functional mechanisms through which the electricity is realized. However, developments in science cannot be limited to the schemes in which technological systems work in everyday situations. As the ends of science cannot be reduced to how it is practiced, scientific practices cannot be reduced to how their objects are technically produced. Hence, the phenomenotechnique of scientific practice must not be equated with a particular technical mode of action that realizes its goals by employing technologies.

Science Is Irreducible to Technology

What is the extra element that marks the difference between the working of technologies in everyday practice and how they function within scientific practices? Bachelard's answer to this question is formulated along two lines of argument: on the one hand, he makes a distinction between action and knowledge. On the other hand, he stresses the importance of mathematical thinking in scientific practices and argues that this marks a discontinuity with our ordinary practices. The conjunction of these two arguments is what, according to Bachelard, differentiates the phenomenotechnical nature of scientific practices from the technical mode of action found in industrial practice.

The first line of argumentation is concerned with the distinction between action, which is concerned with closed systems, and knowledge, which is concerned with that which is not part of closed systems. In industrial systems, technological processes are used to create the same product over and over again, and any unexpected results would be considered failures. In mass industry, the goal is to create a large number of products that cannot be distinguished from another, and the coming into being of each individual product can be defined in terms of the mechanism through which it is created. In scientific experiments, this is not the case. While scientific experiments are of course designed to test hypotheses and scientists have expectations concerning the result, this result itself is not part of the experimental design. The outcome of a scientific experiment should add to the existing scientific knowledge in a given domain (i.e., scientific experiments are supposed to generate something new).

Furthermore, the degree of certitude needed in scientific experiments differs from the degree needed for technological constructions. While technological constructions contain certitude on a very general level, they lack certitude on the microlevel that science is concerned with and in which things become increasingly complex (Bachelard 2005a, 177). The end of a closed system is clear, such that the degree of certainty of the technologies involved must be relative to the goal of the system. However, because science does not strive for any particular goal, it cannot neglect how this operation works on a detailed level (i.e., it precisely aims to understand the micro-phenomena that technological systems neglect). In scientific experiments, the working of technologies is based on hypotheses, because the specific outcome of the experiment is uncertain. Thus, while technological systems are principally closed, scientific experiments always entail the possibility that something unexpected happens, thereby creating a continuous openness.

The second primary difference between technological and scientific practices is that twentieth-century science concerns a domain "that is of a mathematical sense before having a phenomenal signification" (Bachelard 2005b, 75). Following the Kantian distinction between phenomenon and noumenon, Bachelard holds that, strictly speaking, scientific phenomena do not have their foundation in sensory experience but are products of mathematical thinking. However, Bachelard gives a twist to the Kantian distinction, which conflicts with Kant's conception of the noumenon (Chimisso 2008, 387). While the noumenon is principally inaccessible for human beings in Kant's view, Bachelard holds the fact that some scientific phenomena, such as electrons, can never be sensory experienced as stable objects do not prevent scientists from having epistemological access to them.

Precisely because the phenomena of twentieth-century physics are mathematically formed and cannot be found in ordinary experience can the rationality of science be warranted. Contrary to the Kantian noumenon, Bachelard's noumenon, which is scientifically constructed, is knowable to the human being and functions as the ground for the coming into being of scientific phenomena. The mathematical construction of a rational plane in which scientific objects appear can be understood as "a noumenology that elucidates a phenomenotechnique through which new phenomena are not simply found but invented, constructed and built from all parts" (Bachelard 2005b, 76). In this specific sense, mathematics also appears as a technique for creating a noumenon in which scientific phenomena appear.

This specific mathematical grounding is the second way in which scientific practices differ from ordinary practices in which technologies are involved. This does of course not imply that technological systems do not involve the use of mathematical reasoning. For example, the product output of an industrial factory may well be mathematically predicted in terms of the number of products the machinery is likely to manufacture in a certain amount of time. However, in contrast with scientific practices, the existence of the product (say, a shoe) is not entirely dependent on this mathematical prediction. In other words, contrary to scientific phenomena, the presence of mathematics is not constitutive of the existence of technical products.

However, the sciences should not be reduced to their mathematical character. While, on the one hand, Bachelard argues that the sciences should rationalize reality to construct scientific phenomena against nature, he holds, on the other hand, that these rationalizations should become real in the sense that they can be empirically observed. It is precisely this empirical realization of rational thought that constitutes the experimental character of twentieth-century science. When the notion of phenomenotechnique is placed in this context, it becomes clear that it cannot be reduced to the stable functioning of the involved technologies, but transcends these material aspects. The structure of the phenomenotechnique works on the basis that a mathematical surface is rationally already there and can only realize scientific phenomena within this surface. Within the phenomenotechnique, the rationality of mathematics constitutes a noumenal plane that is potentially subject to change, and material technologies help ensuring that this plane relatively constant, such that they help reveal the phenomena through which scientific objects are constituted.

4.3 SCIENTIFIC PRACTICE BEYOND PHYSICS

It can be questioned whether Bachelard's analysis of (micro-)physics can be taken as a model for the sciences in general. Does his epistemology allow for attributing a specific rationality to different scientific practices in different disciplines? Put differently, does Bachelard's epistemology allow for a scientific pluralism, or does he consider (micro-)physics the most mature science up to his day that functions as a model of rationality for science as a whole?

Cornelius Borck has argued that the brain, as an object of neuroscientific research, is subject to continuous change, because it is made increasingly complex through the technologies used to investigate it. He claims that "'the brain' must be understood as an artifact in the precise sense of the term Bachelard had in mind when he described modern physics" (2016, 118). How the neurosciences conceive of the brain is dependent on how it appears within the phenomenotechnique. Because of the increasing complexity of the brain that scientists discover in relation to the phenomenotechnique in which it is made present, the neurosciences become a diverse field consisting of several subdisciplines that each work on their own set of problems. This suggests a continuity between the natural and the neurosciences in terms of how their object of research is created and how it is made increasingly complex. But to what extent does a continuity in terms of how scientific objects come into being also lead to a continuity in terms of the nature of these objects?

Different Ways of Rationalizing the Real

Bachelard is well aware that the different sciences deal with different objects but does not consider this to be in conflict with his epistemology, because the different sciences each have their own *regional rationalism* (Bachelard 1949a). Because the sciences should aim for making their objects increasingly complex, specialization is a natural consequence. Bachelard attempts to develop a positive account of the implication of this specialization and places it in the context of the actual development of the sciences. Specialization is the positive consequence of the non-substantialist character of the sciences, which leads to the increasing complexity of the objects present in scientific research. Precisely because of this increasing complexity, it is necessary that a plurality of specializations must come into being to do in-depth research into each of these phenomena (Bachelard 1951, 13). Each of these specializations develops its own rational structure specific to how its object is made complex.

This view can be illustrated with reference to the development of the cognitive neurosciences. When the brain is treated as an explanatory substance for cognitive functioning, it is maintained that particular forms of human behavior are the consequences of the interactions on a neuronal level, which in turn may be caused by interactions on a cellular level. Inevitably so, this makes the mechanisms underlying cognitive functioning increasingly complex. To begin to understand this system, one would need knowledge of cell biology, physics, and chemistry. As a consequence, specialized groups are formed that deal with the different aspects of human cognition through the lens of their specific fields. Hence, through its increasing complexity, cognition becomes a scientific object that is scattered across different scientific disciplines that each deal with it in a specific manner.

This complexity causes a plurality of separate scientific disciplines that each offer their own specific perspective on human cognition. To function *as* specialized, each discipline must construct "its" human cognition in a specific way, and they cannot do so without using technologies. When observing the cellular structure of brain tissue through a microscope, a different object comes into being than when studying networks of neurons through the use of fMRI. Each of these different ways of observation comes with different problems, questions, and theories concerning its specific domain. Following Bachelard, one can say that each of these domains constructs its objects within its own phenomenotechnique.

In Bachelard's view, this falling into pieces of cognition into different specific scientific objects does not give rise to the fear that these pieces can never be reconnected into the original object of research. On the contrary, he sees this as a natural consequence of how objectivity is manifested in the sciences; the scattering of the initial object is precisely what transforms it into a truly scientific object of research. He argues that "through scientific thinking, the object acquires a deepened perspective" (Bachelard 1951, 12). The initial object can be described in terms of different conceptual frameworks that each explain its workings on a specific level (cellular, neuronal, etc.). As a consequence, each of these different scientific practices that construct its own objects within distinct phenomenotechnical structures develops its own regional rationalism in relation to the specific technologies it uses and the concepts it employs.

4.4 THE RATIONALITY OF THE SCIENTIFIC PROJECT

On what basis, then, can it be said that the different sciences participate equally in the scientific project? Bachelard searches to answer this question by introducing the idea that the different regional rationalisms in science are unified because they all engage in a rational apodicticity, which ensures "the application of the rational technique of scientific thinking determines a genuine recurrence of rationality" (1949a, 121). This suggests that what the different sciences (e.g., the natural and the life sciences) keep together is the rational techniques they apply to make scientific objects appear within a rational domain.

Bachelard conceptualizes scientific practices in relation to the project of science that searches for the possibility to engage in theoretical activity. The rationality at work within the scientific project should aim for the ongoing correction of previous errors. This suggests that insofar as other sciences participate in this same rationality, there is no a priori reason to make a strict division between physics and the other sciences. No matter what technologies, research objects, or concepts are present, the different sciences can all be considered scientific insofar as they participate in the rationalist project of science at large. In other words, the structure of the particular phenomenotechnique at work within a particular scientific context functions because it participates in a scientific rationality. Accordingly, the micro-activity of scientific practice can only be considered scientific to the extent that it is part of the macrostructure of science as a rational project. Now, what is the relation between the micro-aspects of scientific practice that occur through the phenomenotechnique and the general rationality of the scientific project?

The Scientific City

In Bachelard's view, science, as a project, occurs in a "scientific city," and the inhabitants of this city seem to be perfectly rational actors that come to a collective understanding of a scientific phenomenon, because they all participate in the rationality of science. In Bachelard's view, objectivity and rationality can only emerge out of social exchanges between different scientists. Only within collectives is it possible that individual scientists do not obey their instinctual desires and drifts, which would culminate in an imaginative experience of nature, and are classified by Bachelard as epistemological obstacles that should be overcome (Chimisso 2013, 401).

The idea of the scientific city is grounded in how Bachelard conceives of the actual structure of scientific practice. For example, when discussing that most of the papers appearing in scientific journals are written by several authors, he stresses that this cooperative aspect is typical for how rationality manifests in the *newest* manifestations of science (i.e., in twentieth-century physics) (Bachelard 1951, 9).[5] Collective rationality functions as a model for how science *should* be organized and how scientists can attain a position of rational apodicticity. The idea that rationality can only be obtained when participating in actual scientific practice and by relating to other scientists is most dramatically expressed in Bachelard's posthumously published fragments on poetics, where he writes,

> I am of the conviction that an active rationalism must be associated with scientific labor, transforming all knowledge into scientific knowledge. Thus, if I were to write a new book as a rationalist, I would have to go to the school of one of the contemporary sciences. One can no longer be *rationalist by oneself*, on the fringe of contemporary scientific activity (Emphasis in original.). It is necessary to learn together with the workers of rationality. (Bachelard 1997, 9)

The lonely individual is unable to produce rational knowledge, because rationality is constituted in the scientific city through the discursive practices of scientists. Only in this environment can the human mind be rectified such that it overcomes epistemological obstacles. The individual, on the contrary, is capable only of producing desires and dreams that are private by definition and can only be expressed in poetry. Hence, the social structure is a necessary precondition for engaging in the scientific project (Chimisso 2001, 93).

Scientific Experience Is Rational Experience

Collective rationality not only is limited to personal interactions between scientists but also structures how scientists relate to the phenomena they study. In the scientific city, the collective discursive layer of the written word provides an extra foundation of the sciences, which is as important as the construction of the scientific noumenon through mathematical techniques. Polemically, Bachelard refers to this discursive structure as providing a *bibliomenon* that provides an ongoing stability for the existence of scientific objects within the scientific community. Furthermore, through the constant interaction with the scientific phenomena that exist within the bibliomenon, scientists remain constantly aware that the objects that science concerns are not natural objects but phenomena that are artificially created. Hence, this discursive structure functions as a mechanism that helps scientists to continue to overcome the epistemological obstacles they are faced with (Bachelard 1951, 8). This continuous wake-up call is especially important for Bachelard, because the danger of falling back into the nonrational mode of everyday life remains ever-present.

According to Bachelard, the collective and discursive structure of the scientific city guides how science is practiced on a microlevel. The micro-experience of scientific practice participates in the scientific project that guides all scientific thinking. No matter which technologies, research objects, and concepts are present, the different sciences can all be considered scientific insofar as they participate in the rationalist project of science at large. In other words, the structure of a particular phenomenotechnique at work within a particular scientific practice only functions insofar as it participates in a scientific rationality; the micro-activity of scientific practice can only be performed to the extent that it is part of the macro-aspect of science as a rational project. Only within this rational structure, which is materially represented in the scientific city, can scientists relate to *scientific* objects (Bachelard 1949a, 121). The structure is realized in the construction of a rational environment in which scientists are capable of continuously overcoming epistemological obstacles in their interactions with each other and the books in which scientific objects are stabilized. Accordingly, how the scientific city is organized is contingent on the internal development of scientific rationality and should avoid having any inference with external factors, because these function as obstacles to the very possibility of scientific rationality.

4.5 PHENOMENOTECHNIQUE AS PHENOMENOTECHNOLOGY

Bachelard primarily understands technologies as techniques that are involved in the construction of a rational environment in which science occurs. In the end, the most important use of technologies is how they help stabilizing scientific phenomena in experimental environments. The micro-functioning of technologies in scientific experiments is only relevant insofar as it contributes to science in a general sense as a rational project. The macro-aspect of the scientific project is thus prioritized over the micro-experience of scientific practice, which is why Bachelard eventually disconnects all materialities of scientific practice from science itself. In this section, I reconceptualize Bachelard's notion of phenomenotechnique as *phenomenotechnology*. Introducing this concept in a postphenomenological framework helps emphasizing that the intentional relation between scientists and the world is mediated by a phenomenotechnological structure rather than by isolated technologies. At the same time, precisely because I understand a phenomenotechnology in phenomenological terms, this does not require assuming a form of rationality beforehand. Rather, what is taken to be "rational" is itself shaped within the mediated relation between scientists and the reality they investigate.

The Limits of Bachelard's Phenomenotechnique

As demonstrated above, in Bachelard's view, the lonely individual is unable to produce rational knowledge, because rationality is constituted in the scientific city. Only in this environment can the human mind be rectified such that it overcomes epistemological obstacles (i.e., relate to objects that are reified within science). Hence, the social structure in which it becomes possible to engage in a state of rational apodicticity is a necessary precondition for engaging in the scientific project.

How then must Bachelard's notion of phenomenotechnique be adjusted to incorporate the role of technologies in scientific practice? The idea of science as crucially dependent on techniques in Bachelard's work seems to have at least three different meanings. Firstly, "technique" refers to the technologies through which scientific objects come into being. Secondly, "technique" refers to the mathematical creation of a noumenon in which these phenomena must appear. Thirdly, "technique" refers to the creation of the scientific city, in which scientific phenomena are rationally discussed among scientists. Thus, despite Bachelard's (and many others) interchangeable use of the words "technique" and "technology," the "technique" in phenomenotechnique should be understood as allowing for phenomena to be *made* and for the *making* of a rational environment in which the sciences can conduct objective research.

On this view, technologies are primarily conceptualized in terms of allowing for the creating of a stability within which scientific objects appear. After the creation of a phenomenon within a phenomenotechnical structure, the phenomenotechnique disappears the moment that the phenomenon is stabilized. Instead of remaining part of the materiality of scientific practice, the scientific phenomenon becomes part of the rationality of the scientific project that transcends all practices. While Bachelard attempts to integrate the role of technologies in epistemology, he is primarily interested in how they function as techniques to let scientific phenomena come into being in a rational manner. Ensuring the rationality of the scientific project is prioritized over the micro-experience of scientific practice, which is why Bachelard eventually disconnects all materialities of scientific practice from the scientific project. In other words, when the rational is realized, the scientific phenomenon and the technologies that accompany it dissolve into the rationality of the scientific project.

Phenomenotechnology: A Postphenomenological Reading of "Phenomenotechnique"

Then, how is Bachelard's epistemology still relevant to debates in—and between—the philosophy of science and science studies? Does a manufactured rational environment provide a link between the practical and theoretical aspects of scientific practices? When considering that Bachelard conceptualizes the working of technologies in terms of creating stability across scientific experiments, these ensure that scientific practices are rationally organized through this stability. In other words, because of this stability, different scientists are capable of theorizing about the same scientific phenomena. However, the theorizing that occurs within the rational structure of science is itself independent of how scientific phenomena are created. It presupposes a social structure (the scientific city) and a mathematical structure that must already be there to ensure the rationality of both the phenomena appearing in the phenomenotechnique and the theories that are subsequently developed about these phenomena.

When comparing this view of Bachelard's with how technologies in scientific practice are dealt with in postphenomenology, its focus on how technologies, as part of the phenomenotechnique, make phenomena more complex allows for gaining a deeper understanding of the "scientific" aspects of scientific practice. While Don Ihde primarily focuses on how technologies make phenomena visible by detecting signals that cannot be perceived with human biological capacities, Bachelard's epistemology adds to this idea that scientists must also make sense of these phenomena in a conceptual manner. Framed in terms of Ihde's notion of *material hermeneutics*, a postphenomenological epistemology primarily focuses on how the phenomena that technologies detect are transformed into readable images that are read at a glance by the observing scientists. Ihde locates the hermeneutic moment in this process in the transformation of what the technology has detected into an image that is brought within the domain of human perception.

As I hope to have shown in this chapter, with Bachelard, this approach can be augmented by also focusing on how scientific phenomena arise as the consequence of (i) concepts that already frame how a phenomenon comes into being and (ii) how these concepts play an integral part in constituting the meaning attached to the phenomenon (i.e., how it is constituted as a scientific object). Considering this, hermeneutics in science is not limited to how technologies transform what they have detected into readable images, or bring scientific phenomena within the human sensory domain in another way but also occurs in the conceptual (trans)formation of scientific objects. Hence, there is a doubled hermeneutic at work in scientific practice: on the one hand, it concerns how material technologies make a phenomenon perceptively visible, while on the other hand, it also concerns how this visible phenomenon is conceptually transformed into a relevant *scientific* phenomenon that can be meaningfully reasoned about in scientific practice.

Let me illustrate this double hermeneutics with an example of research in the cognitive neurosciences. When a researcher is interested in studying visual attention in this scientific domain, she typically encounters a brain scan that is meant to explain the neuronal mechanism that correlates with this type of behavior.[6] This includes the transformation of what is detected through fMRI into a brain scan that enters the human perceptive field, as well as the transformation of the pattern depicted on this scan into the scientific object of visual attention. This second transformation occurs because an image of the brain entering a human perceptive field does not entail on its own that the content of this image can be related to visual attention. To put it bluntly, a picture of the brain is not necessarily a picture of cognitive functioning. There must be a second hermeneutic moment that allows for the transformation of an image of the brain into an image of visual attention.

A second advantage of integrating the work of Bachelard into a postphenomenological epistemology is that it allows an understanding of how technologies influence how scientific phenomena are actively created. In postphenomenology, technologies are centralized in this process. Don Ihde terms them "epistemology engines" that open up new domains of scientific inquiry by creating new paradigms within which scientific research occurs (Ihde and Selinger 2004). For example, he holds that, today, fMRI necessitates understanding human cognition in terms of brain activity. According to Ihde, because scientists can only access the brain in their relationship with mediating technologies and can never step out of this mediation, they can only perceptively encounter the brain through fMRI. The Bachelardian notion of phenomenotechnique makes clear that the cognitive neurosciences do not deal with the object "brain" but with the increasing complexity of human cognition that can be understood in terms of neuronal activity.

Bachelard's notion of phenomenotechnique does not isolate individual technologies from the broader structure and aims of scientific practices. In postphenomenological terms, one can say that the relation between scientist and the reality that they investigate is mediated by a *phenomenotechnology*, rather than by "concrete technologies." Replacing the term "phenomenotechnique" with "phenomenotechnology" is not mere wordplay: while "technique" refers to the structure within which scientific objects are constructed, "technology"—in a postphenomenological vocabulary—must be understood in phenomenological terms. Accordingly, a phenomenotechnical structure is "phenomenotechnological" because it structures the intentional relation between scientists and the reality they investigate. Within this structure, then, the scientific character of this reality comes into being as a consequence of the specific way in which scientific objects are realized.

4.6 CONCLUSION: EXPLAINING INSTEAD OF ASSUMING THE EPISTEMOLOGICAL RUPTURE

The postphenomenological reformulation of "phenomenotechnique" as "phenomenotechnology" makes the epistemological rupture that Bachelard simply posits a subject of inquiry: as indicated, Bachelard understands technologies as part of the phenomenotechnique that helps creating an epistemological rupture with our ordinary experience. Only in this domain can rationalist activity take place that ensures the coming into existence of objective knowledge. In the end, it is the rationality of science per se that is capable of discovering new scientific phenomena that can instantiate theoretical change, and not the interactions between humans and technologies that constitute new phenomena scientists can theorize about. However, his epistemology does not offer an answer to the question of how scientific objects come into being within the practical interactions between scientists and technologies and the relationships between scientists.

For Bachelard, scientific practice continues to be discussed in terms of the preconditions necessary for engaging in the scientific project, thereby reducing practical interactions to be of interest insofar as they are a manifestation of a form of rationality internal to scientific practices. As I suggested above, when understood as a phenomenotechnology, there is no need to posit an a priori form of rationality that structures scientific research. When understanding the phenomenotechnique as mediating the reality that scientists study (i.e., as phenomenotechnology), it becomes possible to both consider the benefits of Bachelard's approach and allow for the possibility to make the constitution of rationality a subject of phenomenological investigation. In other words, it becomes possible to turn Bachelard's epistemological rupture into a subject of investigation by considering how the reality of science—as rationalized—takes shape as mediated by a particular phenomenotechnology.

In the next chapter, Bruno Latour's empirical philosophy is presented as an important attempt to study the coming into being of the epistemological rupture. He presents a radical alternative to Bachelard's view: while Bachelard assumes a rationality that is internal to science, Latour holds that rationality is not prior to, but is the outcome of, scientific practices. As such, the rationality of science and the specific shape it takes in relation to scientific instruments and in the relation between scientists can become a subject of empirical investigation. As will be shown in the next two chapters, following this route allows for more adequately fulfilling Bachelard's insight that every philosophy of science should depart from an analysis of scientific practice.

NOTES

1. This short detour does not aim to give a full overview of Bachelard's philosophy of imagination. For such an overview, see, for example, Kaplan 1972.

2. Such a psychoanalysis of objective knowledge is the purpose of Bachelard's *The Psychoanalysis of Fire* (1964).

3. An excellent discussion of these concepts can be found in Rheinberger (2010, 25–37).

4. This closely relates to Baird's (2004) conceptualization of thermometers as instances of representational thing knowledge. However, the crucial difference with Baird's line of thinking that I follow is Bachelard's emphasis on how the thermometer opens up new domains of theoretical inquiry rather than that it is a mere stabilization of existing theories of heat. That is, from my perspective, the thermometer is part of a phenomenotechnique that constitutes a new domain of scientific objects.

5. The claim that cooperation between scientists is typical for twentieth-century science can be doubted on historical grounds. For an account of the cooperation in the seventeenth century between scientists (natural philosophers) and the need for experiments to be public events to be observed by other scientists, see, for example, Daston and Galison (2010), Shapin and Schaffer (2011).

6. For a detailed discussion of the neuroscientific study of visual attention and the way it includes both perceptual and conceptual transformation, see chapter 7.

Chapter 5

Bruno Latour and the Difference between Technical and Technological Mediations

INTRODUCTION

How do technologies function within scientific collectives? Is it indeed to be assumed that, as Bachelard did, understanding scientific practices requires asking scientists to "tell us what you think, not when you quit the laboratory, but during the hours when you leave your ordinary life behind you and enter scientific life" (Bachelard 1968, 11)? As the previous chapter suggested, accepting this radical divide between scientific and ordinary practice problematically assumes that the rationality of science is guaranteed by the internal logic of the scientific project. From this perspective, how scientists relate to technologies—as part of the phenomenotechnique—is also explained in terms of a rationality that is external to scientific practices. Would it not be possible to develop a different understanding of scientific practices, performed by scientific collectives that relate to technologies, without presupposing that these practices are rational in themselves?

In the previous chapters, I explored the "scientific" in scientific practices and investigated to what extent human-technology relations are specifically shaped through the theoretical or rational character of the sciences. In this chapter, I focus on the "practical" part of scientific practice. The point of departure of my analysis of the practical character of the scientific, and the roles of technologies in it, is the work of the French philosopher-anthropologist Bruno Latour, whose work can be understood as a radical alternative to Bachelard's internalist conception of scientific rationality.

5.1 UNDERSTANDING SCIENCE AS PRACTICE AND UNDERSTANDING PRACTICE AS SCIENCE

As the previous chapter outlined, Gaston Bachelard traces the working of scientific instruments in scientific practice back to a scientific rationality that eventually remains detached from any practical context. He postulates a rationality that is internal to science and in which scientific practices participate. Bruno Latour's philosophy can be considered a radical alternative to this view: he holds that rationality is not prior to, but the outcome of, scientific practices. Latour thus denies that there is an essential discontinuity between the reality of science and the reality of everyday life that can be attributed to the rationality of science. Accordingly, it becomes possible to explore the working of technologies in scientific practice without letting them dissolve in a scientific rationality and to highlight the dependency of scientific practice on the presence of technologies.

Whereas Bachelard is primarily interested in how scientific instruments help maintain the presence of a scientific rationality, Latour, by denying the a priori presence of scientific rationality, attributes a much wider variety of functions and meanings to them. He attempts to understand scientific instruments in their own right without explaining them away with reference to the rationality of the sciences. The importance attributed to scientific instruments becomes clear from the following statement by Latour and Woolgar: "The observation of radioactivity is entirely dependent on the [gamma] counter" (1986, 64). In principle, this claim can still be taken to mean that the reality of radioactivity is constructed by how the gamma counter rationalizes its phenomenal appearance. However, as will become clear in this chapter, Latour takes this statement to mean something different. Instead of attributing the construction of the reality of radioactivity to scientific rationality, Latour understands this process in terms of the specific efforts and skills of the scientists who use to the gamma counter, skills and efforts which, as demonstrated below, cannot be explained in terms of a detached scientific rationality.

Ready-Made Science and Science-in-the-making

At the heart of Latour's approach to science is his refusal to attribute a special kind of rationality to the sciences that would distinguish them from other human affairs. Contrary to Bachelard, he does not presuppose any epistemological break between ordinary and scientific practices that can explain the rationality of the latter. Bachelard argues for the importance of close analyses of scientific practices to challenge existing philosophical systems, but he primarily examined the scientific objects that are the end product of scientific practices. This is precisely the view of science that Latour criticizes when distinguishing between *ready-made-science* and *science-in-the-making* (1987, 4). He argues that science has a Janus-faced nature. The left side of Janus says that when scientific facts are ready-made, they appear as unquestionable, undisputable properties of nature discovered by the proper rational method. However, the right side of Janus speaks when the facts are under construction and when it is undetermined what nature is and what is the proper rational method to access it. According to Latour, this latter process cannot be interpreted as guided by rationality of nature, because the meanings of terms such as "rational" are precisely what is at stake in science-in-the-making.

Latour argues that the reason why science has a Janus face is that established scientific facts are *black boxes*. Because of this, "no matter how controversial their history, how complex their inner workings, how large the commercial or academic networks that hold them in place, only their input and output count" (Ibid. 3). However, black boxes can only function as such when remaining closed, and this is what happens when examining ready-made science. When opening black boxes (i.e., when examining science-in-the-making), it becomes clear that the facts of science are the product of controversies and that this process cannot be interpreted as guided by rationality or nature. In science-in-the-making, it is impossible to distinguish between context and content, such that the outcome of a controversy cannot be understood without reference to the context in which it occurred. Thus, the closing of controversies occurs within the messy relations between scientists, instruments, institutional concerns, and personal interests.

Scientists often make an appeal to nature as ultimately settling controversies. However, according to Latour, it is not yet determined what nature is in processes of science-in-the-making, such that it cannot be an explanatory factor when considering how controversies are solved. Just as science itself, nature is Janus-faced: "Nature is the final cause of the settlement of all controversies, *once controversies are settled*, [but] as long as they last *Nature will appear simply as the final consequence of the controversies*" (Ibid. 98) (Emphasis in original). From the perspective of science-in-the-making, nature is only there after disagreement was made impossible, because no possible competitors can gather a sufficient amount of resources to allow them to continue their battle. In other words, nature, as the outcome of scientific practice, was never already there but is what is constructed in science-in-the-making.

Phenomenotechnique and Science-in-the-making

The idea that nature is never already there but is always what is constructed in scientific practice is a view that makes it seem that Latour's treatment of science is close to Bachelard's. In *Laboratory Life*, Latour and Woolgar quote Bachelard approvingly when discussing how it is through practical and material arrangements that the phenomena of science are constituted. Consider the following quote:

> The central importance of [. . .] material arrangement[s] is that none of the phenomena "about which" participants talk could exist without [them]. Without a bioassay, for example, a substance could not be said to exist. The bioassay is not merely a means of obtaining some independently given entity; the bioassay constitutes the construction of the substance. [. . .] The artificial reality, which participants describe in terms of an objective entity, has *in fact* been constructed by the use of inscription devices. Such a reality, which Bachelard (1953) terms the "phenomenotechnique," takes on the appearance of a phenomenon by virtue of its construction through material techniques. (Latour and Woolgar 1986, 64, my emphasis)

Although Bachelard is quoted approvingly in the above fragment, it makes clear the difference between Bachelard and Latour. For Bachelard, it would not make sense to speak about an objective entity that *in fact* is created through a phenomenotechnique. In Bachelard's view, it is evident that objective entities are created within a phenomenotechnique, because it is for this very reason that these entities can be called objective. Only by creating an artificial reality against nature can we speak of "objectivity" in the first place; the presence of a phenomenotechnique is a prerequisite for the construction of *scientific* phenomena in the first place.

When arguing that scientific entities are *in fact* artificially constructed within a phenomenotechnique, Latour and Woolgar juxtapose objectivity with construction. The separation of construction and objectivity implies a perspective suggesting that while the researchers in the laboratory *act as if* they are speaking about an entity that is objectively out there, they are *in fact* speaking about something that is constructed and does not exist outside of the structure of the laboratory and hence is not objectively out there. According to Latour and Woolgar, one should better understand "objective entities" as created through a material arrangement that provides "inscriptions which can be used to write papers or to make points in the literature on the basis of a transformation of established arguments into items of apparatus" (Ibid. 66). Thus, the phenomenotechnique is part of what Latour sees when opening the black box of a scientific fact; it disappears from view when examining ready-made science, while it has been of crucial importance when examining science-in-the-making.

This different usage of "phenomenotechnique" can be traced back to the different ways in which Bachelard and Latour think about the difference between scientific and ordinary practices. Bachelard considers scientific practices in terms of the rationalization of the real and of the phenomenotechnique as a prerequisite for both the rationality of science and the possibility for empirical observation in science. The phenomenotechnique realizes an epistemological break with ordinary practice, thereby giving rise to the rationality and objectivity of the sciences. According to Latour, however, the rupture between science and everyday life and the supposed rationality and irrationality accompanying these different situations cannot be assumed but must be explained. Latour's phenomenotechnique should be understood as an *instrument* offering inscriptions in a visual form. In doing so, the phenomenotechnique contributes to the coming into being of the black box by constructing inscriptions that narrow down the possibility of disagreement. In sum, whereas for Bachelard, the phenomenotechnique is a medium *through* which rationality and objectivity can enter scientific practice, Latour uses the term to explain their *constitution*.

5.2 LATOUR'S "CRITIQUE" OF CRITIQUE: HOW TO AVOID A METALANGUAGE?

Should Latour's refusal to attribute a special kind of rationality to scientific practices be considered a critique on the scientific enterprise at large? Yes and no. What he does criticize is the idea, found, for example, in Bachelard's epistemology, that there are a priori reasons to distinguish between scientific and nonscientific practices. Latour's criticism of Bachelard can be connected to Latour's idea that no metalanguage or rationality should be used to account for the differences between scientific and everyday practices.

Enlightenment Critique as Denying Hybridity: How Science Adds to Reality

To understand why Latour refuses to speak about science through any metalanguage and still believes that there *is* a language to speak about science, it is necessary to understand how he situates the value of his own work through a discussion around the notion of "Critique." As Graham Harman has remarked, Latour's most important "criticism" of critique is that it makes things less real, while its goal is to make things more real (Harman 2004, 120). This loss of reality is, according to Latour, grounded in what he calls the Modern Constitution: the idea that things and objects are either natural or social, that they either belong to the purity of science or to the messiness of the social world (Latour 1993b, 35–38). However, as Latour painstakingly attempts to show, entities can never be placed in either of the domains.

They are hybrids, continuously oscillating between the two domains, and can never be explained in terms of either of them.

In Latour's view, the hybrids created in scientific practices *add* meaning to reality, often in the form of factual statements. Latour clarifies this process of addition with reference to Whitehead's discussion of the bifurcation of nature with reference to how scientific explanations increase the possibilities in which sunlight can be described: "For us the red glow of the sunset should be as much part of nature as are the molecules and electric waves by which men of science would explain the phenomenon" (Whitehead 1920, 28–29 op. cit. Latour 2004, 244). From this perspective, reality has been augmented through scientific developments, rather than having eliminated other possible relations with the sun.

When nature is not understood as bifurcated, there is no need to assume that reality can be divided into a domain that is "social" and "experiential" and one that is "scientific" and "rational." This position does not necessarily force one to become a relativist with regard to the validity of scientific knowledge but rather circumvents debates between scientific realists and relativists/social constructivists. Latour clarifies this by referring to the complexity of the entities dealt with in the sciences that is insufficiently recognized in philosophical debates concerning scientific realism:

> Philosophy never deals with the sort of beings we in science studies have dealt with. And that's why the debates between realism and relativism never go anywhere. As Ian Hacking [in *The Social Construction of What?* (1999)] has recently shown, the engagement of a rock in philosophical talk is utterly different if you take a banal rock to make your point (usually to lapidate a passing relativist!) or if you take, for instance, dolomite, as he has done so beautifully.[1] (Latour 2004, 234)

When taking scientific objects such as dolomites and human cognition as a point of departure, it becomes immediately clear that science deals with complex entities about which it can be unproblematically said that they are constructed, contrary to about, for example, ordinary objects such as rocks and pieces of wood. This is not to say that these objects do not "really" exist but rather to emphasize that for these things to exist, much work must be done. For example, to observe the cognitive function of human beings, scientific instruments must be used, participants must be gathered and rightly prepared, an experimental setup must be carefully managed, statistical analyses must be performed, images must be discussed with colleagues, and after all this has been done, a neuroscientist can maybe finally claim to have "seen" a relevant scientific object. The end product, the scientific object, is at the end of a chain of previous relations that have allowed it to come into being. This implies that in scientific practice, knowledge and reality are simultaneously coproduced (De Vries 2016, 48).

This relational conception of scientific phenomena resembles Bachelard's notion of phenomenotechnique, which expresses that the existence of scientific phenomena depends on their relationship with scientists, experimental setups, technologies, and earlier scientific theories. However, as demonstrated above, while Bachelard assumes a scientific rationality providing structure to the phenomenotechnique, no such structuring principle is present in Latour's work. The relations within the phenomenotechnique must be mapped and interpreted in terms of their own relationality without reference to an external explanatory principle. Each relation adds something to reality: a new inscription, a new ally, a new relation, a new possible meaning.

5.3 THE CONSTRUCTION OF SCIENTIFIC ENTITIES: PASTEUR'S MICROBES

In *The Pasteurization of France*, Latour states that he aims to "follow the mechanisms that may allow me to understand at once the *content* of a science and its *context*" (Latour 1988b, 12) (Emphasis in original). Pasteur's "discovery" of microbes is a key case study in Latour's oeuvre. The question he searches to answer through this study is how a new entity was added to reality through which it became clear that "there are more of us than we thought" (Latour 1988b, 35). Latour does not attribute the coming into being of this new entity to either the brilliance of Pasteur, who had finally discovered what had always been there, or as the outcome of political and social forces that had a specific interest in Pasteur's work and would benefit from microbes to enter society. Indeed, if content and context really merge, we can no longer set up a dichotomy neither between the "scientific" and the "social," nor between "internal" and "external" explanations of scientific discoveries.

Debunking the Modern Constitution: Pasteur's Extended Laboratory

According to Latour, the false dichotomies of the Modern Constitution are grounded in a mistaken understanding of what a laboratory comprises. An internal explanation of the coming into existence of microbes would be something like this: Pasteur makes a discovery in his laboratory, which is subsequently transmitted to society, thereby adding an extra element to it. In an external explanation, social and political forces actively create interest in Pasteur's work, and it is these forces that are responsible for welcoming the microbe into society. This would imply that microbes were either created *in* Pasteur's laboratory or that they were created *outside* of it. For Latour, both of these explanations are mistaken. The former reduces scientific practice to ready-made science, while the latter is unable to see that the interest in microbes is "the consequence and not a cause of Pasteur's efforts to translate what they want or what he makes them want" (Latour 1983, 144). Latour attempts to escape these either/or explanations by showing that Pasteur's laboratory itself is a place in which the social context and the scientific content of his work merge.

Instead of stable places of "splendid isolation," laboratories must be understood as unstable processes that must constantly be able to determine their own relevance. In doing so, they destabilize the interests of the others they wish to convince. In contrast with Bachelard, then, there is no form of rationality present in the laboratory—there is no such thing as Bachelard's scientific city—that legitimizes its existence. Latour presents the story of Pasteur and the discovery of microbes by discussing the interrelation between the interests of the hygienists, the solving of practical problems in society, and the experiments of Pasteur and his cooperators. While his discussion ultimately focuses on the success of Pasteur, I focus primarily on the questions of how microbes came into being and how they were recognized as a new element of reality. In doing so, I approach Latour's work as an account of how a new object to which human beings can relate is added to reality.

In the second half of the nineteenth century, France was suffering from a contagious disease stemming from an unknown source that was referred to as "morbid spontaneity." This disease was characterized by an enormous uncertainty, and it occurred in almost every element of society. Most of the people traveling to foreign countries came back healthy; most of the midwives did not kill their clients' babies; most of the wounds of people did not get infected; most cattle remained healthy and could be used for farming. However, in some cases, the opposite occurred, causing sickness, death, and rotten portions of food needed to be thrown away. In the absence of a plausible explanation for these sudden tragedies, this disease was treated as appearing spontaneously (Latour 1988b, 32–34).

Several scientists were studying the cause of this disease in their laboratories, Pasteur being one among them. Without the presence of any regularities in its appearance, it seemed almost impossible to understand this disease in terms of a cause instantiating an effect in all these cases. Yet, this was exactly what occurred when Pasteur and his cooperators started to understand all of the irregularities in terms of microbes. However, microbes could only be observed in a controlled environment, which greatly diverged from the variety of circumstances in which people got sick. When microbes were indeed an extra layer of reality, Pasteur had to show not only that he could observe microbes in his controlled environment but also that they existed and played the important role Pasteur ascribed to them outside of this controlled domain. Only in this way could an extra invisible element be added to reality that allowed the explanation of both the cause of "morbid spontaneity" and the variety of occasions in which it occurred.

Latour argues that it can only be understood how microbes became part of our reality when giving up the strict distinction between the inner domain of the laboratory and the outside domain of the society. Initially, microbes did only exist in the laboratory, because studying the properties of microbes required transporting them from the conditions in which they caused diseases. Pasteur needed another environment that "was an *ideal* one for the microbe, since for the first time since the existence of microbes in the world they were allowed to develop *alone* (Emphasis in original). It was also an 'ideal' condition for the observer, since in developing so blithely, the microbe, freed from the competition of other living beings, *made itself visible* by increasing and multiplying" (Ibid. 63). However, it is only when microbes are considered to exist everywhere, rather than only under specific laboratory conditions, that an extra element is added to reality. In Latour's words, "A laboratory microbe is not yet a 'contagious ferment'" (Ibid. 63).

But how to understand the relation between the microbes in Pasteur's laboratory and the contagious ferments that were causing sickness? In one way or another, Pasteur had to clarify that he was speaking about the same thing in these two situations. Latour describes this in terms of the creation of an equivalence between a controlled laboratory environment and the messy circumstances in which people get sick. The equivalence of the two is not a priori given, but Pasteur actively must show that microbes act similarly under different circumstances. This creation of equivalence can be readily observed in how Pasteur linked his microbes to the existence of the so-called anthrax disease. This disease caused the death of a large part of French cattle, giving rise to great concern among French farmers and veterinarians (Latour 1983, 144). As a form of "morbid spontaneity," the anthrax disease had a variety in which no clear pattern could be detected, and it was thought that the disease defied explanation in terms of a single cause.

Yet, proposing this single cause explanation based on his laboratory studies was exactly what Pasteur would do. But how did he succeed in convincing others that microbes not only existed in his laboratory but also caused the death of cattle on farm sites? According to Latour, because Pasteur constantly showed the relevance of his laboratory in a variety of places, farmers and veterinarians were forced to move through Pasteur's work when attempting to cure the anthrax disease. In other words, Pasteur actively showed that his laboratory and the farm sites were indeed equivalent places when microbes were concerned and that his laboratory could *change* the circumstances of the farm site by managing the behavior of microbes. Thus, he showed (i) that his laboratory was of interest to the farmers and (ii) that only in the laboratory could a position of domination over microbes be maintained.

When Pasteur solved the problems of the farms, he needed to show not only that he could dominate the anthrax disease in his laboratory but also that his experiments could help prevent the outbreak of disease at farm sites. "Here, the problem is to find a compromise that extends Pasteur's laboratory far enough—so that the vaccination can be repeated and work—but which is still acceptable to the farming representatives so that it is seen as an extension of the lab science outside" (Ibid. 151). Again, an equivalence between the farm sites and Pasteur's laboratory had to be created, ensuring that the same microbes did the same things in two previously different situations. When this equivalence is created, the farm site has become a laboratory, and only because of this is it became possible to show that microbes are the cause of the anthrax disease. Only by extending the laboratory to the farm sites can a situation in which "wherever the veterinarian comes the small parasite [microbe] has to go" (Ibid. 154) come into being. At this point, it is no longer possible to distinguish between what is laboratory and what is not or between what is scientific and what is social.

Latour's purpose when explaining Pasteur's success in this way is to debunk the myth of the isolated scientist, who, through his participation in a transcendent rationality, can discover what has always been there, on the one hand, and the idea that scientific discoveries are *entirely* dependent on social forces, on the other. Rather, Pasteur's microbes became real not because they were suddenly there but because they interacted with other aspects of the world in which they played a crucial explanatory role. It is only because of their connections with other entities that microbes act in a certain context, and it is only in this context that microbes can display themselves. Only in actively constructing the existence of the microbe did it become possible to ascribe certain actions to them, and only by constructing equivalence was Pasteur capable of showing that the same element acted in two different situations.

5.4 SCIENTIFIC INSTRUMENTS AS INSCRIPTION DEVICES AND THE CONSTITUTION OF NEW ENTITIES

In Latour's story about Pasteur's microbes, scientific instruments and the material context in which scientists work play only a minor role. Although the notion of "laboratory" suggests the existence of a place in which scientists talk with each other, make observations using a variety of instruments, and carefully develop an environment in which phenomena can be controlled,

Latour seems to use it primarily in a metaphorical sense. In other words, his narrative needs a laboratory in which microbes did exist for the Pasteurians because they can be isolated and visualized, in contrast with the farm sites, where the microbes remained invisible. Latour suggests at several places that there is something specific to laboratories that allows for establishing a specific relation with phenomena, yet—for reasons described above—he refuses to call this relation "scientific." However, he does single out something that is *specific* to laboratories when explaining what Pasteur did in his laboratory: "Why did Pasteur gain strength in the laboratory? He did so because there, as in every laboratory, phenomena are finally made smaller than the group of men who can then dominate them" (Latour 1988a, 74). Whereas at the farm site, the entire environment is dominated by the invisible microbes, the laboratory creates a new environment in which the tables are turned; where the microbe has become visible to a scientist who relates to it through a microscope.

Latour highlights as the central force of the laboratory and its equipment that through it, microbes are objectified and individualized because their behavior in specific circumstances and reactions to changes in these can be carefully monitored. The power of the laboratory strongly depends on scientific instruments that Latour conceptualizes as *inscription devices*. An inscription device is "any set-up, no matter what its size, nature and cost, that provides a visual display of any sort in a scientific text" (Latour 1987, 68), and "the final end product of all these inscription devices is always a written trace that makes the perceptive judgment of the others *simpler*" (Latour 1983, 161). Through these devices, Pasteur can manage the behavior of the phenomena under study:

> They are experts inside their own walls at setting up trials and instruments so that the invisible actors—which they call microbes—show their moves and development in pictures that even a child would see them. The invisible becomes visible and the "thing" becomes a written trace they can read at will as if it were a text. (Ibid. 163)

This notion of manageability is further specified in the idea that inscription devices not only organize perceptions but also create an *immutable mobility*. Because of this, inscriptions can travel from one place to another without being subject to change. Latour exemplifies this with an example taken from the history of cartography. In the eighteenth century, the French naval officer Jean-François de la Pérouse traveled through the Pacific to bring back a better map of the location of Pacific islands for Louis XVI. As Louis XVI himself did not travel by sea, his knowledge of Pacific topography depended upon what De la Pérouse brought back to him, something that should remain stable between the Pacific and the Versailles. The only way to do so was by creating a map—that is, an *immutable mobile*—through which something new was inscribed that was stabilized and could move without being destabilized (Latour 1986, 7). The spatiotemporal distance between the place where the inscription was developed and the place where it acted was overcome by this immutable mobility. While Latour primarily discusses this in terms of cartography, we can see a similar process at work when scientists graphically visualize their experimental findings, which readers encounter in scientific papers.

Scientific Instruments beyond Inscription Devices: How Do Instruments Add to Reality?

Does this analysis of laboratories as sites where instruments are built that function as inscription devices exhaustively explain how scientific instruments contribute to the coming into being of new scientific objects? Of course, the behavior of phenomena must be carefully managed to become interpretable, but understanding this management process does not explain why and how a specific kind of phenomenon is studied. For example, Latour makes perfectly clear how the microbes where transferred from Pasteur's laboratory to the environment of the farm sites and shows how this environment changes with the introduction of microbes. However, he does not pay attention to how microbes came into being in the first place or why the Pasteurians would make them a candidate cause for "morbid spontaneity." Because of this, how scientific instruments allowed the Pasteurians (and eventually also farmers and French society in general) to relate to the world in a new way with the introduction of the concept "microbe" remains unanalyzed. Accordingly, Latour's analysis does not capture why microbes, rather than something else, were *added* to reality.

Latour's lack of attention to how the possibility arose to interpret the epidemics in French society in terms of microbes is one criticism made by Simon Schaffer in his review of *The Pasteurization of France*:

> Latour systematically understates the work of experiment. Experimental labor is neglected in *Les Microbes* [The original French version of *The Pasteurization of France*] because the book's narrative is marked by the heresy of hylozoism, an attribution of purpose, will and life to inanimate matter, and of human interests to the nonhuman. (1992, 182)

Schaffer uses Latour's interest in the nonhuman as a critique of his supposed lack of interest in the human, and especially as a lack of interest in the relation between Pasteur's and other scientists' attempting to work on a cure for the French epidemics. However, Latour's "hylozoism" can also be read as an attempt to extend the domain that is of interest for a philosophy of science that takes the role of technologies seriously.

What Schaffer interprets as hylozoism—the anthromorphization of inanimate matter—has a less problematic connotation if the way Latour speaks about the "acts" of microbes is linked to his idea that nonhumans and humans always are part of networks in which they have a relation to one another. For example, when microbes were seen as responsible for the sick cattle, veterinarians and farmers had to engage in different practices than they had before. The network between cattle, veterinarians, and farmers changed with the introduction of microbes; it is in this sense that they should be understood as actors—or as *actants*, in Latour's terminology. From this perspective, it is not needed to say that the microbes have the will to change practices, do so intentionally, or possess certain interests in different practices but simply that their coming into existence instantiated change. The situation after the introduction of microbes is different from the previous one, and this is how Latour understands the "acts" of microbes.

When action is understood in this minimal sense, no specific human characteristics have to be attributed to nonhumans such as microbes and other material devices. However, the revolutionary move that Latour makes is that he also refuses to attribute a specific form of agency, will, or freedom to human actors. He proposes a *symmetric* understanding of humans and nonhumans (e.g., Latour 1987, 144). This symmetry is reflected in how he interprets both humans and nonhumans as "actants." The consequences of this equal treatment of nonhumans and humans are not explicitly worked out in Latour's analyses of scientific fact-making but forms an integral part of how he analyzes the role of artifacts in our society. Not only microbes but also mundane artifacts such as chairs, tables, and hammers can be understood as actants. For example, the presence of a speed bump creates a different traffic situation than a perfectly asphalted flat highway would. In this brief example, it becomes already clear how these different artifacts (perfectly flat asphalt vs. a speed bump) act, in the sense that they instantiate specific practices. While the speed bump invites drivers to slow down, the flat asphalt invites drivers to speed up, thereby creating two entirely different traffic situations in which different rules apply.

Can this analysis be expanded to how artifacts and instruments function within a scientific context as instantiating new *scientific* practices, which open up new ways of *acting* in the world? In Latour's understanding of scientific instruments as inscription devices, which may include technologies, a prior existing framework that must be *inscribed* is already assumed. As a result, a Latourian analysis does not address how the content of the inscription itself is formed. However, the inscribed is also not already there but must be actively worked upon. For example, a brain scan depicting the relevant brain networks for a specific human cognitive task assumes that scientists (i) already understand what "cognition" is, which is (ii) accessed in relation to a specific technology. The interaction between these two elements in the creation of an interpretational framework to understand human behavior is absent in Latour's work, yet it must be understood if we wish to understand how scientists, and subsequently also other human actors, acquaint themselves with a specific understanding of reality.

In the remainder of this chapter, I hope to show that nonhumans can also be understood as actants in a relevant sense in the creation of a specific understanding of reality. This requires a reconceptualization of the Latourian understanding of science-in-the-making, which must be able to account for how nonhuman actants in general, and technologies in particular, allow for the instantiation of *new* practices, in which scientists can develop a *new* understanding of reality. Below, I suggest that a merger of a Latour's semiotic approach and a postphenomenological approach that focuses on how technologies make scientists encounter new experiential phenomena can help doing so.

5.5 SCIENCE-IN-THE-MAKING RECONSIDERED

Earlier in this chapter, I introduced Latour's idea of science as being Janus-faced: the left side of the Janus face speaks in the name of ready-made science, while the other side speaks in the name of science-in-the-making. On the one hand, scientific facts are unquestionable when ready-made and then become heralded for being exemplary of a specific form of scientific rationality. On the other hand, however, when examining science-in-the-making, in which facts are not yet there, the rational method to be used when doing scientific research is at stake. This discrepancy explains Latour's interest in analyzing how facts and scientific objects are constructed in science-in-the-making. But when and how is science being made?

According to Latour, answering this question requires to return to the moment before a clear distinction between content and context can be made (Latour 1987, 5). Readymade science speaks about facts that embody a uniform content that is sharply distinguished from the context in which they have developed. As we saw, in Latour's terminology, when confronted with ready-made science, we are confronted with *black boxes*. However, in science-in-the-making, these boxes are most of the time not closed yet. This forces to ask the following question: "How are we going to account for the closing of the boxes, because they do, after all, close up?" (Ibid. 7). Science is being made before the black boxes are closed, and in this stage, there is no clear distinction between what is factual and what is not.

Latour's approach presupposes the existence of a specific black box, of which the process of making can be retrospectively analyzed. For example, it requires that the existence of Pasteur's microbes is no longer questioned to start to understand the conditions by which they became unquestionable. On Latour's view, the coming into being of microbes can be explained in terms of them being part of a strong network that made it impossible to question that infection and diseases were caused by invisible microbes. Understanding how microbes came into being thus requires deconstructing the network in which microbes are stabilized and unraveling the relation between the actants that hold the network in place. In this section, I show how this idea is grounded in a semiotic take on scientific practice, which, so I will argue, prevent Latour from clarifying how the entity "microbe" emerges in the relation between scientists and the instruments they use.

Latour's Semiotic Approach and Its Limits

Earlier, it was already briefly mentioned that semiotics is Latour's primary entry point into science-in-the-making. Semiotics allows to shake off the ballast of readymade science and to erase the vocabulary that is entrenched in the language of already established scientific facts when studying science-in-the-making. To do so, the analyst of science-in-the-making has to pretend that (s) he does not yet know what her world is made of. This provides an opportunity rather than that it poses a problem:

> The fact that we do not know in advance what the world is made up of is not a reason for refusing to make a start, because *other* storytellers seem to know and are constantly defining the actors that surround them—what they want, what causes them, and the ways in which they can be weakened and linked together (Emphasis in original). (Latour 1988b, 10)

Semiotics is a vehicle to understand science-in-the-making from the perspective of the actors participating in it, instead of understanding scientific practices from an outside point of view that has already attributed a specific structure to reality: "Semiotics is the ethnomethodology of texts. Like ethnomethodology, it helps replace the analyst's restricted and limited vocabulary by the actor's activity at world-making" (Latour 1993a, 131). In Latour's view, semiotics reveals how the actors within practices relate to one another without relying on the presuppositions of the analyst of those practices (Høstaker 2005, 7).

According to Latour, only semiotics is capable of analyzing the transformations of relations between actants, because, he holds, it is the only methodology that does not a priori distinguish between humans and nonhumans.

Hence, only when practicing semiotics can we analyze how nonhumans transform relations within networks and influence human possibilities to act. Central to this methodology is the idea that human subjects can manipulate their relations with objects at will without being changed by objects themselves is false. Both humans and nonhumans perform actions and transform the relations between parts of the network, and both have the possibility to reshuffle possibilities of action. Thus, semiotics should be understood as a methodological continuation of Latour's symmetry principle that does not prioritize the actions of humans over those of nonhumans.

Two concepts from semiotics specifically inform Latour's research: (i) the interdefinition of actors and (ii) the chains of translations (or mediations) within which acts take place (Latour 1988a, 11). *Interdefinition* is a term developed in the work of the French-Lithuanian linguist Algirdas Greimas, which is used to describe how particular actions situate themselves within a context, and that this particular situating allows one to understand the relation between actor and context. Interdefinitions show how actors are shaped by the network they are part of. This suggests a primacy of the networks over actors, but by adding the idea of *translation*, Latour makes clear that there is no stable network that shapes actions. Networks are continuously shaped by the interdefinitions of the actors. The coming into being of scientific objects is, from this perspective, not the discovery of a stable entity, but the creation of stability through continuous movement.

Examining interdefinitions and translations allows Latour to understand how an actor-network is stabilized, dependent on the already acquired stability of the network in which a heterogeneity of different actants can be shown to have transformed each other in their mutual relations. This semiotic approach is fruitful insofar as one is interested in what generates the stability of networks when this stability is not to be attributed to factors that are external to the network. This is what Latour calls the distinctive feature of his interest in networks, which came to be known as ANT: "Whereas, for the sociologists of the social, the great virtue of appeals to society is that they offer this long lasting stability on a plate and for free, our school views stability as exactly what has to be explained by appealing to costly and demanding means" (Latour 2005, 35).

In other words, the central question for Latour is how networks attain stability if we can no longer pretend that stable external factors such as "social" or "rational" exist that fulfill this task (e.g., Latour 2005; Law 2007). Semiotics is specifically invoked to answer this question, as it allows the analyst to enter the world of the actors and the networks that they are part of without arriving with his own presuppositions. In the language of *Science in Action*, a semiotic approach to science-in-the-making is required to understand both the existence and the credibility of ready-made science. In other words, there should first be the existing stable framework of ready-made science that needs explanation. Actor-networks are explanatory devices that allow for retrospectively understanding how specific utterances have been constructed as facts and how specific entities have been constructed as signifiers.

The interpretation of science-in-the-making in terms of ready-made science is why scientific practices are primarily understood as fact-making practices. Latour takes the current existing order as a starting point and aims to reveal the fluid networks through which this order appears. But why is it to be assumed that the actions performed in scientific practices are (only) structured around fact construction? Furthermore, if science adds new entities to the world to which human beings can relate, why not attempt to understand how scientists can start to relate to something new? These novelties will indeed transform networks and come into being within a specific network, but their birth need not necessarily coincide with the birth of a potential new fact. While Latour interprets scientific practices as participating in processes in which black boxes are *closed*, another way to examine them is by understanding them as processes in which new interpretations of reality are *opened*. In other words, scientific practices can be interpreted as practices of *emergence* as well.

Understanding Science-in-the-Making Hermeneutically

When investigating science-in-the-making, Latour takes the external reference frame of ready-made science that needs explanation as the starting point of his analysis. However, when examining scientific practices as practices in which new unstable objects emerge, it is no longer possible to point to a stabilized referent that needs explanation. From this perspective, scientific practice seems not to be completely reducible to how its end products come into being and become part of reality.

The phenomena that scientists relate to do not come into being spontaneously but emerge within observational and experimental circumstances—which can be called a phenomenotechnology. In such circumstances, a relation between an observer and a phenomenon is established. It is in this relation that the phenomena that are potential candidates for becoming facts are being made. Constructing new phenomena and establishing a new relation with them is a central aspect of scientific practice. In doing so, scientists are capable of disclosing the world in a new way and henceforth developing new ways to interpret reality. The active creation of new phenomena in scientific practices that allow for the emergence of a new relation with reality is what I take to be the process of science-in-the-making. Thus, while Latour is interested in revealing the networks that are "behind" what is already there, a hermeneutic study of science-in-the-making is also interested in how something new can emerge in scientific practice.

When interpreting science-in-the-making in terms of the emergence of new phenomena, we must ask which processes underlie emergence. In other words, the key question is how a relation between scientist and phenomenon comes into being in scientific practice. Only when a scientist relates to something that appears as real can we say that (s)he has observed this phenomenon; when considering the emergence of phenomena, the question to be asked is thus how does a relation between a scientist and something that is revealed as both new and real come into being?

The emergence of new phenomena in scientific practice is also the product of collective work that involves both humans and nonhumans. After all, scientists need technologies to relate to the phenomena they study, and the existence of a laboratory in which these phenomena are collectively investigated must be presupposed. In other words, the emergence of new phenomena also occurs within a network-like structure in which humans and nonhumans transform each other, thereby allowing for the construction of new types of observers and new phenomena that can be observed. When changing the network, new relations between observer and observed are made through which new interpretations of reality can be developed. Thus, the construction of new phenomena implies (i) a reshuffling of the relations between observer and observed and, accordingly, (ii) the potential for interpreting reality in a new way.

To what extent can Latour's semiotic approach help us to understand how new phenomena and interpretations emerge in scientific practice? Introducing something new in a network implies that something is added that was not already there (i.e., it implies that reality is disclosed in such a way that a novel element has come into being). As I have argued, the construction of new phenomena should be understood in terms of reshuffling the relationship between observer and observed and opening up potential new ways of interpretation. On the one hand, this seems to imply a symmetry between human and nonhuman, because both observer and observed are transformed through the specific change in relations in a network. On the other hand, there seems to be an asymmetry involved, when the construction of something new implies the interpreting for something new *as new*. Thus, a specific appropriation of the observed is involved in the construction of a phenomenon that is unique to the human component of the relation between observer and observed.

If the construction of new phenomena is dependent on the interpretation and appropriation of the observed, the notion of network must be reworked such that it can account for the active transformation of the act of appropriation. Acts of appropriation occur within laboratories and are dependent on technologies and collectives of scientists and take place against the background of existing theoretical frameworks (i.e., they are dependent on networks of humans and nonhumans). However, we should not mistake dependency for determinacy. Within networks, phenomena are constructed and become visible to the observer. However, attributing meaning to phenomena (i.e., observing phenomena) involves an activity on the side of the observer that is absent on the side of the observed. In other words, when science-in-the-making is understood as a practice of emergence, Latour's symmetry principle cannot be strictly applied, since it requires the attribution of interpretative qualities to humans while not to nonhumans.

The asymmetry in the observer-observed relation consists of the fact that the observer (i) perceptually experiences and (ii) attributes meaning to the observed. Hence, studying science-in-the-making requires studying these endeavors of the scientist in order to understand how interpretations both arise within and constitute new networks. In other words, a study of science-in-the-making should be a study of how actors are reshuffled within networks and how these reshufflings give rise to specific types of interpretations above others. In the next section, I attempt to show how the framework of postphenomenology can function as a starting point in doing so.

5.6 INTEGRATING POSTPHENOMENOLOGY: A HERMENEUTICS OF SCIENTIFIC INSTRUMENTS

Latour and postphenomenology each approach scientific practices differently. A central target of Latour's project of analyzing science-in-the-making is dismantling the Modern Constitution that gives rise to the idea that the world is divided into human subjects, endowed with agency and intentionality, and nonhuman objects, which are mute entities governed by natural laws or human intentions. According to Latour, this worldview is deeply mistaken, because in fact, stable entities that are dividable into humans and nonhumans do not exist, only *hybrid* entities (Latour 1993b, 13–15). He considers Modernity as an extremely effective way to act as if hybridity does not exist and the world can be divided into clearly different domains. He argues for this position along two lines: firstly, he shows how the "Moderns" were capable of setting up the dichotomy between subject and object, which he calls the work of *purification*, and secondly, he argues for the hybridity of reality by showing how practices are always hybrid (are always constituted by both humans and nonhumans, which he calls the work of *mediation*).

The work of purification consists of two complementing moves. On the one hand, nature must be made foreign to humans and be stabilized as something that has always existed. On the other hand, spokespersons of nature must be created who are capable of discovering the secrets of nature and establishing those as facts. As Latour remarks in the context of the vacuum pump of the seventeenth-century natural philosopher Robert Boyle, "despite their artificial construction inside the vacuum pump (such is the phase of mediation or translation), the facts completely escape all human fabrication (such is the phase of purification)" (Ibid. 31). He generalizes this idea as part of a larger schema that guarantees the subject-object dichotomy underlying the worldview of Modernity: (i) humans have constructed nature, yet nature is treated as if it is not; (ii) humans have not constructed society, yet society is treated as if humans did; and (iii) these two domains must remain absolutely different; mediation and purification should not be mixed (Ibid. 32). Thus, although natural phenomena and facts are constructed within laboratories, they should not be recognized as constructions, and although humans simply find themselves in society, they should act as if they have constructed society.

Latour warns to not interpret the creation or the Modern Constitution as a deliberate act created by an evil genius or through a false consciousness or ideology but considers it as the anthropological background of how modern individuals live:

> The relation between the work of purification and that of mediation is not that of conscious and unconscious, formal and informal, language and practice, illusion and reality. I am not claiming that the moderns are unaware of what they do, I am simply saying that what they do—innovate on a large scale in the production of hybrids—is possible only because they steadfastly hold to the absolute dichotomy between the order of Nature and that of Society, a dichotomy which is itself possible only because they never consider the work of purification and the work of mediation together. (Ibid. 40)

The paradox of a constructed, yet non-constructed, nature is, according to Latour, at the core of modern epistemology. Therefore, exposing the Modern Constitution as a myth means simultaneously exposing modern epistemology as mistakenly grounded. In other words, epistemology fails to consider the work of purification and mediation together and only focuses on how a purified subject can establish a decontextualized position from which knowledge about the objects of nature can be obtained. However, when purification is considered an activity instead of a given, it becomes impossible to circumvent the mediation with which it is accompanied.

What is missing when only considering purifications? According to Latour, it is precisely the hybrid nature of the entities in the world that are the product of mediations and how mediations transform practices. Latour locates mediations as acts constituted by "actors endowed with the capacity to translate what they transport, to redefine it, redeploy, and also to betray it" (Ibid. 81). In other words, mediators are actants that change the structure of the network by redefining other actants. Potentially, every actant can have this mediating role. Because this process of changing is often neglected, it is as if science grasps nature through pure reason. For example, when leaving out the story of how Pasteur extended his laboratory to the farmer sites, it seems that it is Pasteur's pure reason, rather than his practice, that allowed him to cure the anthrax disease by linking it to the invisible microbes. However, when considering how mediators function within networks, the disease was solved not because a subject (Pasteur) was finally able to observe an object (the microbe) but because a stable network was present that transformed the farming site into a place where Pasteur could do his work. In other words, linking the anthrax disease to invisible microbes was the product of a series of mediations.

Latour's Analysis of Technical Mediations Applied to Scientific Practice

In *We Have Never Been Modern*, Latour does not yet link the notion of mediation to specific mediating instances. However, in his later studies on the working of artifacts, he explicitly links mediations to specific material entities. He distinguishes between four meanings of mediation: (i) *translation*, (ii) *composition*, (iii) *reversible black-boxing*, and (iv) *delegation* (Latour 1994). Mediating artifacts (i) instantiate new courses of action, (ii) create new types of composite agents, (iii) often conceal the composite nature of agents, and (iv) have programs of action inscribed in them. While Latour establishes these different forms of mediation in the context of the role of artifacts in everyday life, let me attempt to exemplify them with a brief example from scientific practice.

When applying Latour's analyses of mediating artifacts to the technologies that allowed for the visualization of brain activity and thus shaped how human cognition is to be studied, we can see the four types of mediation at work. Imaging technologies (i) instantiate new laboratory practices in which experiments must be developed in which they can be specifically incorporated. They (ii) create a type of researcher who can no longer observe cognition with his naked eyes (e.g., in the form of behavior) but must relate to the imaging technology to be able to relate to the object of research. Imaging technologies (iii) are referred to as methodologies by researchers, thereby presenting them as one among several choices in the study of human cognition and concealing how the cognition is made visible in a different way to the researcher. Furthermore, when communicating experimental results to the public, the work of the scientists is presented in terms of a subject that establishes a relation with a previously unknown object. Finally, these technologies (iv) have inscribed in them that when someone wishes to engage in studying human cognition, he must interpret it in terms of neurophysiology, which can be interpreted as a consequence of the program of action inscribed in the imaging technology.

How a Postphenomenological Understanding of Scientific Instruments Escapes Latour's Critique of Phenomenology

As the above example suggests, also a Latourian analysis already point to the fact that scientific instruments constitute particular possibilities of interpretation. However, how these interpretative possibilities come into being remains under-analyzed in Latour's approach. A postphenomenological perspective, by contrast, is able to also address this aspect of scientific practice. But does a postphenomenological understanding of scientific instruments as mediating the domain in which the objects of research appear reinforces the scheme of epistemology so vigorously contested by Latour? Does it not presuppose a model in which a subject must bridge the gap with an object, and that just the domain in which this bridging problem appears is limited? When approached phenomenologically, this problem can be circumvented: there is nothing to bridge between subject and object, as scientific instruments can be understood as mediating the intentional relation between scientists and the world. Accordingly, not only is the domain in which scientific phenomena appear the consequence of a series of mediations, but how researchers intentionally relate to the phenomena under study is also mediated by the technologies that they are using.

When phenomenologists speak about intentionality, they refer to the fact that when human beings think or have beliefs, these thoughts and beliefs are always about something. Similarly, human beings do not merely experience; they always experience something. Hence, also when scientists have beliefs about an object of research or experience the presence of a phenomenon in observation, these beliefs and experiences are about something. In other words, a researcher experiences or has beliefs about something other than himself. Thus, human intentionality consists of the relation between the human being and the world in which the phenomena (s)he relates to emerge.

Latour is critical of phenomenology and its concept of intentionality and argues that phenomenology is (i) an artifact of the Modern Constitution and (ii) trapped in theories of a bifurcated nature. According to Latour, phenomenology reaffirms the yawning gap between subjects and objects:

> Like so many anxious modernizers, they [the phenomenologists] no longer trace anything but a line between poles that are thus given the greatest importance. Pure objectivity and pure consciousness are missing, but they are nevertheless—indeed, all the more—in place. The "consciousness of something" becomes

> nothing more than a slender footbridge spanning a gradually widening abyss. (Latour 1993b, 58)

Although Latour does not specify which phenomenologists he is targeting, he seems to primarily refer to (a caricature of) the work of Edmund Husserl, who considered intentionality an idea to bridge the gap between the subject and the object pole and to show why human beings can interpret objects in a meaningful way. The notion of intentionality guarantees that human perception is necessarily related to something that is not part of it; it entails the object that is intentionally related to. From the perspective of Latour, however, this position does not solve but rather reaffirms the Modern Constitution, because it presupposes the existence of an isolated subject capable of having a consciousness relation with something distant from it. According to Latour, the concept of intentionality establishes a relation between two poles, which phenomenologists mistakenly take to be separated in the first place.

Latour argues that the consequences of continuing to adhere to the Modern Constitution become most clear in the phenomenological reactions to scientific developments. He uses Whitehead to argue against phenomenologists who would add "to the boring electric waves the rich lived world of the glowing sun" (Latour 2004, 244). This assumes that there are specific qualities that can be attributed to the objects (the "boringness" of electric waves) and other qualities that can be attributed to the experience of the human subject (the rich, lived world of the glowing sun). The reasons to emphasize these distinct human qualities are the fears that the experiences of the human subject will be objectified and that the domain of human experience will be reduced to the language of scientific explanation. Precisely this idea is contested by Latour when he argues that this fear is grounded in the mistaken assumption that something like decontextualized human experience that must be protected does exist. In this way, objects are reduced to the "dull hurrying of nature" that occurs in opposition to the rich inner lives of human individuals (Latour 2008). Following Latour's line of reasoning, phenomenologists are criticized for failing to notice that the objects around us mediate how the world is experienced, observed, and acted in.

No isolated subject exists that is capable of thinking, experiencing, observing, or acting on its own: "I live in the midst of *technical* delegates: I am folded into nonhumans" (Latour 1999a, 189) (Emphasis in original). Actions and objects have an influence on how the world is experienced. There is no escape from the influence of these other actors on how practices are shaped or reality is experienced. Applied to scientific practices, researchers cannot escape the influence of existing concepts and theories, the functioning of the technologies they use, the material infrastructure of the laboratory, or the presuppositions of their colleagues. Researchers are folded into these human and nonhuman elements, as are the objects they study. In other words, how scientists study reality is mediated by the specific way in which they are folded.

As demonstrated above, Latour proposes studying these foldings by "unfolding" the networks that constitute practices. When reconstructing the history of practices (i.e., by studying how practices come into being), Latour shows how our current practices and acts are dependent on the networks of which they are products. Earlier in this chapter, I distinguished between two different ways of interpreting scientific practices. On the one hand, I characterized them (following Latour) as concerned with the closing of black boxes. On the other hand, however, I characterized them as practices of emergence in which new interpretations of reality are opened. To study scientific practices as practices of emergence, it must become clear how scientists relate to the world *within* foldings. Doing so requires the concept of intentionality, yet in a way that it is understood *in terms of the relations between a human observer and the things that allow him to establish a relation with the world.* Postphenomenology is specifically concerned with how combinations of humans and nonhumans establish an intentional relation with reality. Similar to Latour, philosophers such as Don Ihde and Peter-Paul Verbeek argue that no isolated subject exists that establishes a relation with an object but that human intentionality is always shaped in relation to a technological environment (e.g., Ihde 1979; Verbeek 2005). However, contrary to Latour, Ihde and Verbeek hold that the fact that relations to the world are always the consequence of combinations of humans and nonhumans does not require to treat them on entirely equal footing. Human beings are inevitably intertwined with technologies and the other artifacts they use but have capacities that technologies do not. The world made related to through technologies is the world that must be interpreted by the human being. Something distinct in the human being does relate to a technology that makes it possible to intentionally relate to a world.

From this perspective, intentionality is shaped by both the technologies that human beings relate to and the structure of the practice in which these relations occur, and accordingly, the world that is related to is the product of the relation between human and technology within this practical structure. The relation between human being and world is mediated by the technology (s)he relates to, and in this relation, a specific human being and a specific world are constituted. However, as demonstrated above, these relations are not isolated events but occur within practices that are already there and are structured around existing concepts and theories, the relations between scientists, and the specific infrastructure of the laboratory. Integrating the notion of intentionality allows viewing how human beings relate to the world within a folded environment, which can be seen as complementary to the work of Latour, because it allows us to not only understand how specific practices come into being but also how interpretations of reality are opened within this practice.

On the Difference between Technical Mediations and Technological Mediations

For both Latour and postphenomenology, *mediation* is a central concept when describing the relationship between human beings and technologies. However, while Latour speaks of *technical* mediation, postphenomenologists speak of *technological* mediation. However, there are relevant differences between these two notions. Whereas "technical mediation" refers to how previous technical acts that are black-boxed have created a specific practice, "technological mediation" is a concept used to describe how relationships between human beings and reality are shaped *within* practices. Thus, "technical mediations" primarily refer to how the present is constituted in relation to past events, while "technological mediations" shape relations in the here and now.

Latour develops his concept of technical mediation largely independently of his work on the construction of scientific facts, yet he acknowledges that important parallels exist between facts and artifacts: "Technical artifacts are as far from the status of efficiency as scientific facts are from the noble pedestal of objectivity" (Latour 1999a, 193). Just as we should not understand the construction of scientific facts in terms of an assumed underlying rationality, we should not understand technical objects in terms of an assumed efficiency. Technical objects are "never the beginning but the *end* of a long process of proliferating mediators, a process in which all relevant subprograms nested one into another, meet in a 'simple' task" (Ibid. 192). Thus, both in the domain of science and in the domain of technical actions, Latour urges us to understand both of these in terms of not their presumed objectivity or efficiency but the mediating acts that allowed for the attribution of these terms to certain states of affairs in the first place.

In a discussion about Latour's concept of technical mediation, Jeff Kochan has argued that Latour treats mediation as a black box that—although remaining closed itself—serves as an explanatory term to understand other events. He argues that Latour explains the distance between laboratory scientists and the layman in terms of the establishment of black boxes that have an indisputable authority. The central concept in the distancing between laboratory and layman is the concept of mediation: "Mediation is the primitive phenomenon by which all other phenomena will be explained, but which is itself closed to explanation. Latour seems to believe that we have no choice but to accept the inexorable hegemony of mediation" (Kochan 2010, 593). Kochan treats this as a criticism against Latour that reveals the modernist foundations of Latour's explicit non-modernist position, because of his affirmation of technoscientific productions over the interests of human individuals.

In line with the argument developed in this chapter, I treat Latour's use of mediation as a consequence of his interest in the post hoc deconstruction of the networks "behind" scientific factuality and technical efficiency. For example, Pasteur's microbes have been constructed and become a new inhabitant of our society. Therefore, it seems impossible to neglect the "successes" of the mediations of which the microbes are the product. In other words, technical mediations are used to explain ready-made science and are explanatory of how scientific facts and objects are constructed. Latour must assume the existence of scientific facts precisely because he is primarily concerned with how they came into being. This is reflected in his understanding of technologies as instruments, which are solely understood in relation to the construction of the entity whose existence is to be explained.

The idea of *technological* mediation is not limited to this retrospectivity. Because of its grounding in phenomenology, it treats technologies not in relation to the existence of something external (say, a microbe) but as allowing human beings to relate to the world in specific ways. Furthermore, the idea of technological mediation embraces that these relations are never between a decontextualized subject and world but are shaped within a specific practical context. The intentionality of the subject is thus constituted in relation with both the technology and the specificity of the practices in which the subject does relate to a technology. Thus, whereas Latour's concept of technical mediation ends with the closing of black boxes or the construction of specific practices, postphenomenologists—while acknowledging the value of understanding the construction of practices in terms of technical mediations—have a broader interest when speaking about how technologies shape our relation with the world in a specific way. This latter focus also allows to consider scientific practices as practices of emergence and makes possible to investigate how new human-technology relations give rise to the coming into being of new scientific objects.

5.7 CONCLUSION: AN EMPIRICAL PHILOSOPHY OF TECHNOSCIENCE: TOWARD A METHODOLOGICAL BASIS

The reconceptualization of science-in-the-making through the lens of technological mediation requires a methodological shift from a semiotic to a hermeneutic approach. The question of how to understand the role of scientific instruments in scientific practice should be answered through a methodological framework other than the one Latour employs to investigate how scientific facts are constructed and become black boxes. While the latter issue can (at least partly) be addressed in terms of the deconstruction of networks, as it presupposes that a specific fact has already come into being, the former question is not bound to this condition. Therefore, a methodology to study how scientists relate to technologies should allow one to draw conclusions about how scientists relate to technologies in real time, which does not limit their influence on how they help construct scientific facts.

Latour's style of philosophy has been repeatedly referred to as "empirical philosophy" (e.g., De Vries 2016; Latour 2008; Schmidgen 2014), but to what extent is this actually reflected in how he approaches science? Latour himself argues that his philosophy should be considered empirical, because it acknowledges that

> actors know what they do and we have to learn from them not only what they do, but how and why they do it. It is *us*, the social scientists, who lack knowledge of what they do, and not *they* who are missing the explanation of why they are unwittingly manipulated by forces exterior to themselves and known to the social scientist's powerful gaze and method (Emphasis in original). (Latour 1999b, 19)

However, when studying actors in scientific practices, he must understand their actions in relation to something that is not yet empirically witnessable: the constructed fact or entity *as* a black box. This is most clearly visible in his analysis of the coming into existence of Pasteur's microbes, which function as a starting point from which all previous acts are interpreted. This seems to seriously undermine the idea that we should learn from actors about what they do and why they do it. After all, *we*, as analysts, know that something like a microbe (or another scientific object or fact) has come into being; *they*, as actors, do not know that. Hence, there seems to be a clear interpretive surplus on the behalf of the analyst when describing the actions of the Pasteurians.

Is there a similar surplus when studying science-in-the-making hermeneutically? Not necessarily, because the advantage of a hermeneutic approach consists precisely in the fact that neither the scientists nor the philosopher interested in scientific practice know what eventually will be made. This allows to understand the working of technologies not only in relation to future events but also as constituting a specific relation with the world that allows researchers to start to observe new phenomena that might lead to new interpretations of reality.

Will this make the job of the analyst easier? Probably not, because where Latour could structure actions with reference to (at least) one certainty (the constructed fact or entity), a study of science-in-the-making lacks such a starting point. The question to be answered then is not whether the analyst has a certainty from which other events can be interpreted but whether it is possible to discover what the certainties of the researchers under study are: why *they* perform the acts that they perform, and how *their* reality is shaped in relation with technologies, but also in relation with existing theories, concepts, and epistemological frameworks (i.e., the phenomenotechnology). In this way, scientific practices can be studied as scientific practices proper, without a priori equating scientific practices with fact-making practices. In the second part of this book, I develop an empirical approach that *does* allow for studying science-in-the-making and exemplify this approach through a discussion of two case studies from the neurosciences.

NOTE

1. For an analysis of the rhetorical strategies behind the many references to rocks and tables in debates concerning scientific realism, see Edwards et al. (1995).

Part II

A POSTPHENOMENOLOGICAL ETHNOMETHODOLOGY OF NEUROSCIENTIFIC PRACTICE

Chapter 6

Postphenomenology and Ethnomethodology

Studying How Reality Is Accomplished through the Appropriation of Technological Mediations

INTRODUCTION: HOW TO STUDY THE CONSTITUTION OF SCIENTIFIC OBJECTS

At the beginning of the first part of this book, I proposed understanding the process in which scientific objects are made present in scientific practices in terms of the display of a shared epistemic stance within a collective of scientists. In the remainder of that part, I focused on how the mediating role of scientific instruments in processes of object constitution can be understood. In this second part of this book, I focus on how to study empirically how collectives of scientists appropriate these technological mediations. If rationality cannot be attributed to scientific practices beforehand *and* the reality of scientific objects cannot be equated with the technologically mediated perceptions that allow for seeing them, how then must the coming into being of a shared epistemic stance be understood? To address this problematic, the second part of this book is concerned with the following question: How does a shared epistemic stance with regard to the reality of scientific objects come into being within scientific collectives?

Postphenomenological studies of scientific practices have thus far focused primarily on how technologies change observational practices by altering the perceptual field of the observing scientist (Ihde 1998; Rosenberger 2011, 2013). Also, the postphenomenological understanding of hermeneutics in science is strongly visually oriented, as it understands the interpretative flexibility—or multistability—of visual objects in terms of the gestalt quality of their presentation (e.g., Ihde 2009, 56). However, a shared reality is constituted not only through the mediation of the perceptual field of the scientific observer but also by the interpretative shaping of the phenomenon that is to enter one's perceptual field. Accordingly, I suggest understanding the practical accomplishment of reality as the accomplishment of an interpretative space, in which perceptual phenomena become present as objects that are conceptually meaningful. In doing so, I aim to further flesh out Verbeek's (2016) suggestion that studies of technological mediations must be accompanied by a study of how human beings *appropriate* those and how a shared reality is constituted within this process.

To do so, I turn to ethnomethodology (EM). Ethnomethodologists are interested in showing how a shared reality is established through action: action that can be observed empirically. As formulated by Harold Garfinkel, the aim of EM is to make explicit "how members' actual ordinary activities consist of methods to make practical actions, practical circumstances, common sense knowledge of social structures, and practical sociological reasoning analyzable," and aims to do so "'from within' actual settings, as ongoing accomplishments of these settings" (Garfinkel 1967, vii–viii). If science cannot be assumed to participate in a special form of rationality, then an ethnomethodological approach to scientific practice has the potential to reveal the collective forms of action through which scientific objects come into being. In other words, an ethnomethodological perspective opens up the possibility to study how a shared reality comes into being for the practical purpose of *doing science*.

Harvey Sacks, the founder of conversation analysis (CA), worried about the absence of a clear methodological grounding of the ethnomethodological program and proposed investigating the practical actions constituting a shared reality through the analysis of conversations. He aimed to develop a sociological program grounded in observable social activities, modeled after primitive observational sciences, such as nineteenth-century biology (Lynch and Bogen 1994, 66). Practices are, according to Sacks, already well-ordered for the participants in them and therefore accessible to observers without preparation, but they can be analyzed by collecting conversations between participants. The task for the analyst is to collect these observations systematically and make them methodically reportable and reproducible to others. At the same time, this method also applies to the reports of scientists themselves, since "they provide for the reproducibility of their actions on the part of themselves or others, by the use of methods" (Sacks 1992, 804). This view of science opens up the possibility to investigate the methods through which scientists make their actions reproducible and reportable, because this is how Sacks understands the practical possibility of doing science. How CA can be used to study scientific practice is illustrated by discussing two laboratory studies of neuroscientific practices drawing on CA: Michael Lynch's *Art and Artifact in Laboratory Science* (1985) and Morana Alač's *Handling Digital Brains* (2011).

A combination of EM and CA opens up the possibility to study empirically how a shared reality to which scientists can relate to comes into being, but these approaches—as I show in this chapter—problematically assume a distinction between the (supposedly mundane) practice of science and scientific theorizing. This affects how the technologies that are present in laboratories are treated by these approaches, because relations between scientists and technologies are located in the (nontheoretical) domain of practical action. For example, when discussing the distinction between the attitudes of daily life and scientific theorizing, Garfinkel writes, "To avoid misunderstanding I want to stress that the concern here is with the attitude of scientific *theorizing* (Emphasis in original). The attitude that informs the activities of actual scientific inquiry is another matter entirely" (Garfinkel 1967, 272, fn. 8).

The purpose of this chapter is to break through the supposed dichotomy between these two attitudes by developing an approach to the study of scientific practices that embraces the intertwinement of theorizing and the practical concerns of scientific inquiries. As I will show, focusing on the appropriation of technological mediations may aid in this regard. This approach to the study of scientific practices aims to reveal two things: (i) how different technologies mediate forms of reasoning by giving rise to new scientific objects and constituting new norms for what is treated as valid knowledge and (ii) how such appropriations of mediations constitute a meaningful reality within scientific collectives. From this perspective, the practical interactions between scientists and technologies are constitutive of the forms of action that constitute a discursive space, in which scientific objects can come into being and become interpretable.

6.1 EM AND REALITY AS A PRACTICAL ACCOMPLISHMENT

From an ethnomethodological perspective, a shared reality is not so much stable, but rather *actively stabilized*. Harold Garfinkel proposed studying this process of stabilization by considering a shared reality as "an ongoing accomplishment of the concerted activities of daily life, with the ordinary, artful ways of that accomplishment being by members known, used, and taken for granted" (Garfinkel 1967, vii). Accordingly, the "realization of reality" is something that can be ethnomethodologically described. Garfinkel's project can be understood as an attempt to describe the traceable interactions between individuals through which realities are practically accomplished.

Harold Garfinkel and Empirically Investigating the Accomplishment of a Shared Reality

Garfinkel denied that either the meaning of the actions of individuals or the reality that individuals share is the product of following external codes or rules. He holds that there is no order external to practices through which a meaning is given to the events occurring but rather that the reality of specific events and the meaning attributed to them is constituted through the rules to which participants in practices orient. If these rules cannot be explained with reference to a system that is external to actual practices, this prompts the question of in what other way group members can collectively relate to a shared reality. According to Garfinkel, understanding how a shared reality comes into being requires to empirically investigate how members themselves remain oriented to this reality in practice.

In placing direct emphasis on empirically analyzing the empirical interactions between individuals, Garfinkel challenges the idea that intersubjectivity is grounded in a cooperative framework between actors. Instead of being the product of a seemingly deliberate cooperation between actors, intersubjectivity should be understood as always present in social interactions, regardless of the intentional motives of the actors that participate in them. At the beginning of his career, he already noted that

> the big question is not whether actors understand each other or not (Emphasis in original). The fact is that they do understand each other, that they *will* understand each other, but the catch is that they will understand each other regardless of how they *would* be understood. (Garfinkel 1952, 367, op. cit. Heritage 1984, 119)

Garfinkel holds that the incongruence between how actors would like to be understood and how they are actually understood should be taken seriously, so as to remain faithful to the ethnomethodological attempt of "the discovery of common culture [. . .] *from within* the society by social scientists of the existence of common sense knowledge of social structures" (Garfinkel 1967, 76–77) (Emphasis in original). By starting "from within society," Garfinkel departs from the idea that a shared reality between *different* actors can be understood in terms of aligning the perceptions and goals of *individual* actors. Only within an already situated environment can actors develop and pursue their goals and relate to a recognizable field of objects. What people see and relate to thus comes into being only in social practices and in the relation with other actors and requires a shared form of vision within which objects become collectively interpretable. However, "the work involved in the coherence of a phenomenal field" (Garfinkel 2002, 97) is typically surpassed in sociology. When focusing on the member's methods to accomplish a shared reality, Garfinkel aims to bring this work into the center of sociological analysis.

While how Garfinkel analyzed the coming into being of a shared phenomenal field focuses specifically on how objects can be perceived, he believed that a similar process was constitutive of the non-perceptual aspects of a shared reality. Borrowing a term from the German sociologist Karl Mannheim, he holds that every actor always uses a *documentary method of interpretation* when making sense of reality, which "consists of treating an actual appearance as 'the document of,' as 'pointing to,' as 'standing on behalf of' a presupposed underlying pattern" (Garfinkel 1967, 78). The documentary method already applies to the most basic level in which individuals interpret reality. For example, if I walk around a table and occasionally close my eyes, only to see the table again when opening my eyes, I treat these two appearances not as two randomly generated events but as pointing to an underlying pattern, namely, the table I am walking around. The key move that Garfinkel makes is that the underlying pattern ("table") is not an idealization made by an individual consciousness but that assuming this pattern is the product of interactions between individuals. If I were to deviate from treating the table in front of me as a stable entity, for example, by asking, "Is there a pattern underlying the set of appearances of something brown and rectangular in front of me?," others would hold me accountable for this action. Accordingly, I must come up with an account of why I am not sitting down at the table and using it to lean on. For example, I might say that I am pursuing a philosophical question, which may (or may not) be treated by others as a convincing account of my deviation of the interpretative pattern "table."

This example indicates that my perception of the object "table" is structured by the maintenance of an underlying pattern and that this pattern is visible in the interactions with others, because I am held accountable for its ongoing maintenance (Ibid. 9). The case in point here is that approaching the maintenance of a shared reality in this way opens up the possibility of *how* it is constituted within the interactions between individuals. Garfinkel urges sociologists to actively pursue this line of research instead of constructing fictitious examples like the one given above, which remain abstract and artificial.[1] In developing this perspective, Garfinkel thus paved the way for empirically studying how a shared reality is accomplished by focusing on how actors hold each other accountable for the maintenance of an underlying pattern that allows for a continuous flow of practical action.

It is this aspect of Garfinkel's EM that makes it worthwhile to integrate into philosophy of technological mediation; the appropriation of what is revealed through scientific instruments can be analyzed in terms of how this revealing is dependent on particular norms that are oriented to in scientific practice. As is demonstrated below, this allows for the consideration of technologies in what

Garfinkel calls the constitution of Galilean objects (i.e., of the objects that science speaks about). However, making this shift requires first investigating whether an ethnomethodological perspective can indeed be unproblematically extended from the study of everyday life to the study of scientific practice.

6.2 EM AND THE RE-SPECIFICATION OF SCIENCE: THE CONSTITUTION OF GALILEAN OBJECTS

Garfinkel seems to maintain a distinction between the *theory* of science and the *practice* of everyday life. When discussing the differences between the attitudes of daily life and scientific theorizing, Garfinkel states that "scientific rationalities, in fact, occur as stable properties and as sanctionable ideals only in the case of actions governed by the attitude of scientific theorizing. By contrast, actions governed by the attitude of daily life are marked by the specific absence of these rationalities" (Garfinkel 1967, 270). However, Garfinkel argues in his later works that EM has the tools to "respecif[y] the natural sciences as discovering sciences of practical action" (Garfinkel 2002, 278). How do these different perspectives on the possibility to study science ethnomethodologically relate to one another?

The Rationality of Science and the Practical Purposes of Scientific Activity

Garfinkel's early remarks on the rationality of the sciences are made in the context of the relation between the scientific discipline "sociology" and the world of everyday action. He is concerned with the fact that sociologists primarily "study the features and conditions of nonrationality in human conduct. The result is that in most of the available theories of social action and social structure rational actions are assigned residual status" (Garfinkel 1967, 263). Terming certain conduct as "irrational" assumes that (i) there is a rationality in place within sociology and (ii) this sociological form of rationality can be unproblematically transferred to the domain of everyday life. The type of sociology that aimed to identify instances of irrationality by applying rational sociological models later come to be known as a *sociology of error* (e.g., Bloor 1976). When distinguishing between the rationality of scientific theorizing and the rationality of everyday life, Garfinkel aims to show that while a certain rationality is constituted within the discipline of sociology, it is mistaken to assume that this rationality can be unproblematically transferred to the domain of everyday life. In doing so, he aims to challenge the conception of sociology as providing a form of rationality that can be normatively applied to the context of everyday action.

In *The Rational Properties of Scientific and Common Sense Activities*, Garfinkel shows the difference between everyday rationality and scientific rationality by discussing the difference between the presuppositions underlying both. He presents five (cf. Garfinkel 1967, 272–77):

(i) In everyday life, actors assume an undoubted correspondence between the appearances of objects and an underlying object appearing in this particular fashion, and assume that other actors do the same. Scientific rationality, on the contrary, prescribes to suspend the belief that the objects of the world are as they appear.
(ii) In everyday life, the relevant features of objects that stand out are dependent on the plans that actors aim to execute. By contrast, scientific theorizing suspends the judgment with regard to the relevant features of objects, but rather aims to conceive them regardless of their place in the real world.
(iii) In everyday life, actors assume the correctness of a standard time to schedule and coordinate their actions with those of others, of which they assume they adhere to a similar standard time: there is "one big clock identical for all" (Ibid. 274). This aspect is absent in scientific theorizing. In this context, time serves not to coordinate events in relation to others, but to identify cause-effect relationships, regardless of their relation with standard time.
(iv) In everyday life, actors assume that they share a background with others such that when making sense of events, different actors will "code" an event in a roughly similar way. This common intersubjectivity is absent in scientific theorizing. Instead of conceiving of others as experiencing subjects, those are treated as ideal, disembodied beings that make sense of phenomena in an objective manner.
(v) The events of daily life are informed by the assumption that part of the background of others always remains private, that is, inaccessible to other actors yet of significant influence on how they conceive of events. Everyday rationality embraces this assumption and tries to act accordingly. In scientific theorizing this is no longer the case: "All matters that are relevant to [the] depiction of a possible world are public and publicizable" (Ibid. 276).

The case in point for Garfinkel is that while the principles of scientific theorizing do indeed give rise to a scientific rationality, this form of rationality cannot serve as an explanatory scheme for everyday life events. When sociologists consider their task as to identify irrationalities against the models derived from a scientific rationality, they neglect that everyday life has its own form of rationality—one that does not coincide with the rationality of scientific theorizing. For this reason, the theoretical attitude of science must be sharply distinguished from the practical forms of rationality guiding everyday conduct.

The Sciences as Accountable Sciences of Practical Action and Practical Reason

In Garfinkel's early work, the specific characteristics of scientific theorizing serve to juxtapose the rationality of everyday life with the rationality of science. However, in his later work, the rationality of the sciences becomes a topic in its own right, because he begins to treat practices of scientists in the same terms as everyday life practices, namely, in terms of how scientists hold each other accountable for the maintenance of an underlying pattern that allows for a continuous flow of practical action. Applied to this new domain, "ethnomethodology is directed to the reform of technical reason, and doing so with the premier aim of specifying the work of the social sciences and the natural sciences as naturally accountable sciences of practical action and practical reason" (Garfinkel 2002, 93). This should be a central focus of EM, because society is increasingly shaped by rationally formalized scientific theories and models. Understanding these processes of formalization and their foundation in practical action allows to reveal how the *plenum* of everyday reality is transformed into the reality that is relevant to the rationality of the sciences.

The ethnomethodological study of scientific practice goes back to Garfinkel's principle that every utterance and action is made for "all practical purposes" (1967, 7). Hence, understanding how the reality of scientists is constituted requires investigating how scientists do so for all practical purposes (i.e., by focusing on how they shape a meaningful context of interpretation). In line with the general program of EM, this approach must be sharply contrasted with a study of scientific practice by means of interviewing scientists, because the practical purposes at stake in answering an interview question are profoundly different from the context in which the reality of science is shaped. In this context, Garfinkel, Lynch, and Livingston speak of a difference between studies *about* scientific practices and studies *of* them (1981, 132). While the former provides detailed descriptions of the social networks that scientists function, stress their financial stakes, and analyze their social roles, the latter is primarily concerned—as is EM—with what makes the occupation "scientist" significant in the first place, namely, the aim to make valid knowledge claims about the reality that is investigated (Heritage 1984, 298).

Studies of scientific practices in an ethnomethodological sense focus on how scientists maintain and manage an order in the unstable and constant stream of phenomena they encounter and ask what the member's methods are that underlie the maintenance of a specific order. Maintaining this order implies the typification of what enters the phenomenal field into a collectively witnessable object, such that members can hold each other accountable for the reality of it. These methods are not grounded in a scientific rationality that exists external of scientific practices, but must be understood as local and contingent ways to stabilize what enters the phenomenal field. They "are probatively suited to the specific thing's accountable production as workplace-specific, discipline-specific work in ordered shop floor phenomenal field details in-and-as-of-the-thing's-generality—and these NOT analytically by explicating lexical generics but contingently analytic, and therein *just in any actual case*" (Garfinkel 2002, 102) (Emphasis in original). From this perspective, the reality of the objects that are constituted in scientific practices is approached not as grounded in the underlying structural of an external nature but as constituted within the process of maintaining order in the phenomenal field. Uncovering the processes through which this reality is constituted is the core of EM's project to re-specify the work of the social and natural sciences.

The Shop Floor Problem of Scientific Practice

Making encounters with phenomena topically and collectively relevant to the involved parties such that a shared reality is constituted is what Garfinkel calls the *shop floor problem* (Garfinkel 2002, 108–12; Lynch 1985). Understanding the constitution of reality in scientific practice in terms of a shop floor problem by no means implies the relativization of the importance of the sciences or counters the idea that the scientists produce objective knowledge. Rather, the objectivity and rationality of scientific practices are interpreted as processes that can be traced in time rather than participating in transcendental forms of objectivity and rationality. In fact, the shop floor problem of scientific practices is the creation of what Garfinkel calls *Galilean objects*, which are "'Details in structures' stand[ing] proxy for independent Galilean objects, not the lived *in vivo* achieved things of immortal ordinary society. Galilean objects are things independently of the work of achieving them" (Garfinkel 2002, 108, fn. 30). The challenge for an EM of scientific practices is accordingly to re-specify how scientists construe details in the structures they observe into Galilean objects that seemingly transcend the work that gave rise to them.

Identifying details in structures requires the use of *indexical expressions* as a method of identification, which is the in vivo work involved of scientific practices. However, to be able to stand as proxies for Galilean objects, it is precisely the indexicality—the here-and-now-ness—of expressions that must be overcome. To create a generalizability of such objects as appearing outside of the locality of scientific practice, it is necessary to work upon a "programmatically relevant distinction between objective and indexical expressions, and a programmatically relevant substitutability of objective for indexical expressions" (Garfinkel and Sacks 1986, 166). In scientific practice, indexicality is constituted as objectivity. Within this act, objects come into being that seemingly transcend the situated character of scientific practice. This is what EM identifies as the shop floor problem of scientific practice, which can be studied in terms of the norms that scientists hold each other accountable for.

A primary contemporary figure proposing an ethnomethodological approach to study scientific practices is the American sociologist Michael Lynch. According to Lynch,

> The ideals of objectivity and scientific method can be identified as mythopoetic constructs made in the service of actions that in their own domain are ordinary. Again, this does not imply that methods are necessarily faulty or that it makes no sense to speak of objective states of affairs. Although no transcendental grounding may warrant its efficacy and certainty for all time, nothing precludes scientific conduct from being orderly, stable, reproducible, reliable, and ordinary. (Lynch 1997, 151)

Lynch proposes that EM should aim to *re-specify* the central epistemological topics of scientific practices, such as "representation," "measurement," "proof," and so on. Here, "a 'respecification' of these topics [does not] mean a [. . .] redefinition of the meaning of terms but a way of investigating the different activities in which 'order,' 'logic,' 'meaning,' and so forth are locally and practically relevant" (Lynch 1993, xi, fn. 1). In doing so, *epistemological* issues are transformed into *epistemic* concerns (i.e., instead of understanding scientific practices as concerned with issues that can be logically solved by means of theory, Lynch proposes to treat these as belonging to the domain of practical action).

In line with this proposal, Lynch approaches judgments such as "(adequate) representation" and "(adequate) measurement" as indexical judgments that scientists hold each other accountable for and that are practically relevant terms through which the reality of the object under study is constituted. The epistemic categories that scientists use when shaping the reality of their object must be empirically determined and may not coincide with notions such as "measurement" and "representation" that have been the central categories in the theory-oriented philosophy of science. Accordingly, one of the aims of his approach to scientific practice is to reveal which epistemic categories are practically relevant and how the use of the categories constitutes norms to which scientists orient themselves during the constitution of the reality of scientific objects.

The Rational Realization of the Real as Practical Accomplishment

What does such a perspective on the constitution of reality within scientific practice imply for our question of the relation between the "ordinary" and the "theoretical" aspects of scientific practice? In Lynch's ethnomethodological account of scientific practice, scientific theorizing cannot be explained in terms of a (set of) contemplating cogito(s) external to the ordinary course of practical actions. This does not imply that scientific theorizing can be reduced to practical actions nor does it eliminate rationality from the domain of science. What an ethnomethodological approach to scientific practices instead points to is that the creation of the "Galilean objects" of science requires the integration of observable details in structures that are indexically present into more complex scientific objects. This requires a concept of practice in which the making present of theoretical objects in the rationality of practical actions can be accounted for. Accordingly, making sense of the here-and-now-ness of observable effects in scientific practice requires attributing meaning to them in terms of what is not empirically observable in scientific practices. This process can be analyzed in terms of how scientists hold each other accountable for *how* observable details in structures are made to stand as "proxies" of scientific objects.

Understanding the rationality of scientific theorizing in terms of constituting the reality of scientific objects in scientific practice allows the connection to an ethnomethodological understanding of scientific practice with the Bachelardian understanding of scientific objects as realized within a phenomenotechnique. Also, in the phenomenotechnique, the practical interactions scientists have with technologies and each other are entangled with and structured by the integration of different complex observable phenomena into scientific concepts, and vice versa. However, while Bachelard's analysis of scientific practice in terms of a phenomenotechnique is based on an external assessment and can, to a certain extent, be considered an idealization of the rationality of scientific practice, EM offers the tools to analyze "from within" the processes through which scientific objects are rationally realized.

6.3 INVESTIGATING SCIENTIFIC PRACTICES AND THE APPROPRIATION OF TECHNOLOGICAL MEDIATIONS THROUGH CA

Understanding the practical validation of epistemic judgments requires a focus on two interrelated aspects: on the one hand, one should focus on the work through which a meaningful reality is practically accomplished, and on the other hand, one must understand how epistemic judgments are made correspondent with this reality. Both of these aspects can be considered by studying the conversations between scientists in which the results of their experiments are discussed, as well as by analyzing how these discussions are shaped by the technologies mediating how scientific phenomena become present in scientific practices. Postphenomenology has emphasized the actively mediating role of technologies in this process but has tended to overlook how human beings subsequently understand and act upon these mediations. Therefore, Verbeek has recently argued that analyses of technological mediations deserve to be complemented with analyzing human *appropriations* (2016, 191).

To also address how technological mediations are appropriated, I suggest augmenting postphenomenology with insights from CA, as applied within ethnomethodological research. Related to EM, CA studies, from a participant's perspective, how people use language as a medium to construct a reality in which meaningful actions can be performed (cf. Lynch 1985, 1993; Potter 1996; Sacks 1992; Te Molder and Potter 2005). The goal of integrating CA into an ethnomethodological perspective on scientific practices is to reveal the epistemic norms that scientists hold each other accountable for when making knowledge claims. In holding each other accountable for certain epistemic norms the reality of the object under study is shaped. The presence of these norms can be observed in conversations between scientists and developed in the relation between scientists and the particular technologies they use.

In this section, I aim to show two things: (i) how the accomplishment of reality and the practical validation of epistemic judgments can be understood as discursive (i.e., conversational and gestural work) and (ii) how making specific this work can be understood as making visible how technological mediations are appropriated within scientific practices.

Empirically Studying Laboratory Practices (I): Lynch's EM Study of Psychobiology

Among the first wave of laboratory studies conducted in the end of the 1970s and the early 1980s, Michael Lynch's *Art and Artifact in Laboratory Science* was explicitly inspired by EM. Lynch aimed to treat neuroscientific practices and the objects arising within them as phenomena that are worth studying in themselves and that should not be treated as helping to settle existing philosophical debates: "Treating the contingent productions of scientific work as evidence for principled epistemological positions and arguments holds the work accountable to issues and questions which are not demonstrably part of the practical setting in an invariant fashion" (Lynch 1985, 275). In doing so, Lynch wishes to move beyond philosophical debates on whether we should adopt a "realist" or "anti-realist" stance against unobservable theoretical entities or whether scientific facts depict a stable external world or must be understood as constructed through social interests. Rather, the order established within the practice of doing science is the topic of investigation. To do so, the established facts and objects should not be subject of study, but Lynch instead aimed to analyze "from within" scientific practices what where the accountable concerns that scientists attribute to one another in the flow of scientific conduct.

Lynch's study is situated at a university-based psychobiology laboratory primarily studying neurotransmitters in both the human and the animal brain, and the regenerative capacity of the brain, also known as brain plasticity. Studying the practices of these neuroscientists from within requires being able to study their activities and how they hold each other accountable for these activities. To do so, Lynch proposes using CA, which allows the study of "talk which accompanies the work as that work is underway; not talk about the work but talk in the work, talk which is part of the work" (Ibid. 10). Drawing on the ethnomethodological tradition and its interest in studying practices in terms of the specific *shop floor problem* that is dealt with, Lynch study focuses on how the measurable data is read by scientists and how specific readings of this data allows for the ongoing continuation of their work in an orderly manner.

Crucial in Lynch's approach is the idea that scientific objects and their treatment cannot be cut loose from the project that temporalize, and thereby order, the work conducted in scientific practices. It is by orienting themselves to a project that the actions of scientific collectives are structured and within which certain features of the obtained data stand out as relevant. However, "the relatedness of actions in a project's sequence of actions [. . .] was not available in the continuous monitoring of the spectacle of events which where witnessed in the lab" (Ibid. 55). Accordingly, "the continuity of a project is not a 'real-time' continuity [. . .] but the analysis [of practical activities] need to supply [. . .] a way of indexing successively displayed actions to their appropriate projects" (Ibid. 68). By analyzing conversations between scientists, Lynch aims to reveal which actions must be performed to realize the continuity of the practices that are oriented to as ordering themes within scientific practices.[2]

Lynch identifies the identification and removal of so-called artifacts in their experimental data as a central shop floor problem of the neuroscientists he studies. Artifacts are interferences in the experimental setup that researchers understand as distortions of the observability of the natural features of the world. These can be caused by technological failures or other circumstances interfering with the naturalized experimental circumstances. An important aspect of neuroscientific practices is to detect these artifacts and find ways to eliminate them from the data under study. This indicates that the obtained data "*are* and *are not* the things studied *per se*; they are conditions within which "the things" are addressed in particular instances of shop work" (Ibid. 10). The obtained data must be actively read and interpreted to make claims about the brain and its functioning. By studying the conversations in which artifacts are eliminated from the data, Lynch aims to show how interpretations of experimental data come into being within scientific collectives, and how specific interpretations allow for perceptual phenomena to "stand proxy" for the scientific object under study.

Focusing on artifacts makes Lynch's study specifically relevant for our understanding of scientific instruments, because artifacts exist by virtue of the technologies used to observe the human brain. The observation of the brain and its functioning is mediated by the electron microscopes used by the scientists that Lynch studies, which instantiates a process in which brains must be aligned with the particular technical specificities of microscopes. Studying the detection of artifacts allows one to bring to the fore the conditions under which scientists speak of a "proper" microscopic observation of the functioning of the brain.

The process of observation starts even before the neuroscientists start to explicitly look through microscopes to observe brain tissue. In the context of a project studying plasticity through observing how a rat's brain recovers from the effects of a lesion, Lynch notes that "the requirement of microscopical observation necessitated the rendering of the animal to comply with the conditions of instrumental observation" (Lynch 1985, 35). As electron micrography does not allow for observing continuous organic processes yet is the only way to observe the brain on an ultrastructural level, this required the "sacrifice" of a large number of animals to display how the brain reacts and recovers from a lesion in a temporally ordered series of displays of the rat's brain.

The relevant questions have now become to what extent the set of displays of the brains of several dead rats can actually stand proxy for the observation of recovery processes in the living brain of a single rat and to what extent the specific way of observing organic processes by means of electron micrography shapes how the functioning of the brain is understood. Lynch's analysis of the detection and treatment of artifacts focuses on the former question and attempts to show how the "making-stand-proxy" of series of displays of dead rats' brains for the recovering capacities of the living brain requires the elimination of artifacts during the process of making the animal comply with the conditions set by electron micrography.

Analyzing the conversations during the identification and treatment of artifact serves to show that studying how scientific claims come into being requires a focus on the technical vocabulary of natural science, instead of being explained in terms of the vocabulary of the social sciences: "The *technical* access to these issues of the social production of natural order was distinct from those technologies of inquiry which are featured in sociology methods texts and departmental pedagogies" (Ibid. 276) (Emphasis in original). How the interpretational frameworks of scientists are shaped by the technologies they use is not at stake in Lynch's study, but he instead focuses on showing how artifacts become accountable phenomena that problematize the visibility of natural phenomena. At stake is thus not primarily showing *how* the objects of interest in the neuroscientific laboratory come into being, but rather *that* this process of coming into existence cannot be reduced to the social scientific vocabulary that is entrenched in explanations in terms of institutional networks, or social and personal interests. When studied from within, so Lynch holds, such explanations appear as abstractions from how order is actually produced in scientific practices. Scientists hold each other accountable not in terms of institutional networks or social interests but within the domain to which their technical vocabulary applies.

However, as can be inferred from the above example in which organic process are made to comply with the conditions of electron micrography, studying the practices of neuroscientists allows to reveal how the way technologies are appropriated by scientists gives rise to specific conceptualizations of biological processes. Although this is not Lynch's focus, I suggest that explicitly focusing on how technological mediations are appropriated helps making clear how scientific technologies give rise to new typifications of observed phenomena and embody interpretational frameworks accordingly. In Lynch's words, artifacts "provided circumstances in which the *positive* manner of addressing a field of instrumentally revealed phenomena was interrupted" (Ibid. 84). So, clearly, Lynch's work suggests that a combination of CA and EM cannot only be used to study how order is obtained within scientific practices but can also serve as a starting point for understanding the extent to which technologies shape how scientists interpret reality. When researchers encounter artifacts, they are confronted with troubling interferences with the phenomena they aim to describe and interpret, thereby having to actively distinguish between "real" and "unreal" aspects of instrumentally mediated perception.

Empirically Studying Laboratory Practices (II): Alač's Study of fMRI

How scientists attribute meaning to specific technologies during the observation of scientific phenomena is explored in more detail in Morana Alač's recent (2011) study on how fMRI scans are used to make claims about the human brain. Building on the tradition of laboratory studies, Alač aims to show how the reality constituted in relation with fMRI scans is the product of multimodal interactions between scientists and between scientists and brain scans in the laboratory. According to Alač, among the most important things thus far neglected in laboratory studies is that the visibility of the objects that scientists investigate "is relative to the circumstance of their practical and multimodal engagement" (Ibid. 18). She specifically aims to reveal the importance of gestural interactions in the production of fMRI visuals by showing that they are grounded in the "lived world of the hands" (Ibid. 158). Drawing on videotaped material, she aims to show how human meaning making in the neurosciences is multimodal and importantly concerns gestural interactions.

Lynch's work is of specific interest to Alač, because his focus on how artifacts are eliminated in neuroscientific practices points to the assumption in neuroscientific practice that "the moving body must become invisible [. . .], so that the 'embodied' mind (i.e., the brain) can be studied" (Ibid. 69). However, her focus is significantly different: whereas Lynch was still primarily concerned with the ethnomethodological question of *order*, Alač's primary focus is on how scientists ascribe *meaning* to the phenomena they encounter in scientific practice. She specifically zooms in on how more experienced neuroscientists teach novices to identify artifacts and how this is accompanied with a specific way of coming to ascribe a meaning to the experimental data that is deemed appropriate.

Alač's way of understanding the processes through which neuroscientists attribute meaning to visuals of the brain is loosely grounded in the semiotics of the American pragmatist Charles Sanders Peirce. Drawing on Peirce, Alač proposes understanding fMRI scans not as images of the brain but instead as diagrammatic signs that can be used to conceptualize the relation between brain areas and the cognitive performance they help to instantiate. From a Peircean perspective, images are signs that have the same simple qualities as the objects they stand proxy for (e.g., one can think of a painted picture of an apple as an image). Diagrams, on the contrary, signify relations between different aspects of objects that can be understood in a variety of manners and allow for experimentation: "Diagrammatic signs show what cannot be observed otherwise while consenting for engagement. This engagement has the potential for generating insights regarding the relationship between the represented elements" (Ibid. 43).

The object of study in the practice of cognitive neuroscientists, human cognition, is not a singular object but is constituted through how different aspects of it are actively integrated through the activity of scientists. Only when understanding fMRI visuals as diagrammatic signs allowing the conceptualization of the complex relationship between activity in the brain and human cognition does it become possible to account for the work of the practitioners that must actively establish links between the different relations signified in the diagram. Without establishing these links, human cognitive functioning cannot be observed, precisely because the "seeing of what cannot be seen relies on the human aptitude to think by exploiting our visual capacity and our skill of handling objects in the world" (Ibid. 28).

The point of Alač's book is that the seeing of cognition can be investigated by studying how cognitive neuroscientists exploit their visual capacity by combining it with gestural interactions through which the relation between the brain and cognitive functioning is made present. This is, for example, illustrated in a discussion of an excerpt of a conversation between an experienced fMRI user and an apprentice learning how the way that visual information is processed in different areas on the brain can be seen on a fMRI scan. The discussion revolves around the identification of a "phase map" of how this processing occurs, which is a visual in which the temporal relationship between the data and experimental stimuli can be detected, and in which the time at which neuronal activity has occurred is indicated with different colors (Ibid. 95). Being able to attribute meaning to a phase map requires observers to be able to see how these different areas in the brain are temporally linked to the stimuli present during the experiment.

In the excerpt, the experienced neuroscientist gradually attempts to teach the apprentice how the temporal relation between experimental stimuli and neuronal processes can be seen. She does so by distinguishing between "seeing" and "looking" and uses "the expression 'actually see' [. . .] [to] refer to the understanding of the data configuration, as opposed to just perceiving colorful images on the computer screen" (Ibid. 98). Next, the experienced neuroscientist starts to use gestures to posit a space between herself and the computer screen, indicating that attributing meaning to the data requires an active involvement of the researcher and cannot be reduced to the properties of the brain scan under study. Using such gestures indicates the potential mutability of the visuals that neuroscientists relate to. Alač uses this as an argument against the Latourian idea that the inscriptions offered by technologies should be interpreted as immutable mobiles, because it shows that within laboratories, the visuals produced by technologies are treated in highly mutable terms: "As seen in the fMRI laboratory, when coordinated with gesturing hands, graphical inscriptions can be quickly transformed but often do not traverse laboratory walls" (Ibid. 100).

In the interaction with diagrammatic signs, so Alač argues, researchers can attribute meaning to brain scans by experimenting with them and stipulating different relational patterns within them through processes of visual and gestural interaction. Through this embodied engagement, researchers slowly become aware of the possible patterns that can be detected in brain scans and become capable of understanding why there is a consensus that some detected patterns are more plausible for understanding the relation between neuronal processes and cognitive function than others. Because she focuses on the gestural interactions in interactions between experienced scientists and apprentices, Alač seems to uncritically assume the ongoing existence of conceptual frameworks of cognitive neuroscience that are instantiated by fMRI. My interest, however, is also in how technological mediations and how they are appropriated give rise to precisely these frameworks. On the one hand, this can be seen as an extension of Alač's study because it is also concerned with how cognitive neuroscientists make sense of the visuals of the human brain they encounter. On the other hand, it breaks with Alač's aims, because meaning making is not considered in terms of knowledge transfers between experts and novices but is rather understood in terms of how the appropriation of technological mediations can give rise to *new* ways of understanding human cognition and the role of the brain in its realization.

6.5 EM AND THE APPROPRIATION OF TECHNOLOGICAL MEDIATIONS

Verbeek's proposal to study how human beings appropriate technological mediations is presented as a corrective to the almost exclusive focus on the workings of technologies in postphenomenology and the neglect of how technological mediations are treated by human beings: "In order to develop a full understanding of processes of mediation, we should not only study 'what *things* do' [. . .], but also how *humans* give meaning to these mediations—both empirically and conceptually" (Verbeek 2016, 190) (Emphasis in original). Indeed, to strengthen the argument that technologies do not determine the constitution of reality by, for example, offering a single perspective on scientific phenomena that determines how scientists interpret the world, it is necessary to reveal the processes of appropriation through which human beings express a certain freedom of interpretation in relation to the technological mediations that they encounter.

A combination of postphenomenology and EM can help study how technological mediations are appropriated in scientific practices. In the context of scientific practice, studying appropriations of technological mediations "makes it possible to investigate how human beings who use scientific instruments give meaning to the mediating roles of technologies and to the mediated realities they help to constitute" (Verbeek 2016, 195). Focusing on how technologically mediated observations and visualizations are translated into the scientific object that is investigated in scientific practices helps to better understand not only how technological mediations co-constitute how scientists act but also how the coming into being of a shared reality is collectively accomplished. For example, thinking of Michael Lynch's study on artifacts, how the brain tissues of dead rats must be rendered in a specific way to stand proxy for organic processes in living animals can be understood as a specific way of giving meaning to how electron micrography mediates our understanding of the brain of an animal. Accordingly, EM can help reveal not only how technologies shape how order is constituted within scientific practices but also how the object of is made present through the appropriation of a technological mediation.

Integrating what I call a *conversational approach to EM* into postphenomenology introduces (at least) two possible pitfalls: (i) by focusing too much on the human interactions through which a shared reality is accomplished, the mediating role of technologies in this process might again be surpassed, and (ii) by reducing scientific practices to the practical action that scientists undertake, the hermeneutic processes within which the conceptual frameworks that also constitute the reality that scientists study are in danger of being neglected. The first pitfall would entail denying what is gained by postphenomenology: taking seriously how the constitution of reality is mediated by the presence of technologies. The second pitfall would conflict with the aim of studying human appropriations to study both *empirically* and *conceptually* how humans give meaning to technological mediations by failing to take seriously that scientific practices have the explicit aim to develop and establish a relationship with Galilean objects (what Heidegger called the relation with the present-at-hand). The latter can only be studied when a hermeneutic perspective is also integrated into EM. By showing how these two pitfalls can be avoided, I aim to clarify how the study of the appropriation of technological mediations differs from traditional ethnomethodological approaches to scientific practices and highlight what can be understood as the *hermeneutic* component of the approach developed in this chapter.

How to Witness the Appropriation of Technological Mediations?

Why is there a risk of overlooking the mediating role of technologies in EM in the first place? Crucial from the perspective of EM is that practices should be explained in terms of the order that is expressed by their members from within yet can be witnessed by an ethnomethodological observer:

> Enacted local practices are not texts which symbolize "meanings" or events. They are in detail identical with themselves, and not representative of something else. The *witnessably* recurrent details of ordinary everyday practices are

> constitutive of their own reality. They are studied in their unmediated details and not as signed enterprises. (Garfinkel 2002, 97, my emphasis)

Order is not to be explained in terms of non-witnessable structures such as "interests," "rationality," or "power-relations" but in terms of the member's methods through which a shared reality is accomplished in concerted action (Garfinkel 1967). The question arising when employing the concept of "technological mediation" is whether and to what extent technological mediations and their appropriations can be "witnessed"?

Let me attempt to answer this question by discussing the second potential pitfall that I identified above: the reduction of scientific practices to the practical actions that scientists undertake. This pitfall can be illustrated by how Garfinkel attempts to re-specify the coming into being of the phenomena of Galilean physics. In line with the ethnomethodological aim to re-specify the natural sciences as sciences of practical action, Garfinkel argues that the phenomena of physics are

> *real-world phenomena* in locally witnessable, naturally accountable coherent substantive, material's details in and as of the things' generality, just in any actual case, *endogenously*, and *that* abbreviations means, without explaining their generality by introducing into the course of the work a generically theorized More that is needed of shop floor details beyond their autochthonous production and accountability (Emphasis in original). (Garfinkel 2002, 264)

Accordingly, the constitution of the phenomena of physics can be investigated "from within" scientific practices by investigating how a series of localized events gives rise to the "Galilean objects" of science that transcend the indexicality of local practices.

Garfinkel attempts to show that the shop floor details and their witnessable accountability constitute the phenomena of physics through a reproduction of Galilei's *inclined plane demonstration of the real motion of free-falling bodies*. Being able to reproduce this experiment has been crucial in the development of modern physics, because only when Galilei would be able to ensure that on every occurrence and if properly conducted by every experimenter, it would be demonstrable that the phenomena he observed transcended the particularities of its situated production. In other words, he had to demonstrate that his results applied not only to some objects in motion in certain settings but to *all* objects that roll (Ibid. 268). This requires the presence of a set of instructed actions describing "all objects that roll" not as Galilean objects but as objects oriented to in the lifeworld that will only behave in the desired manner if their lifeworld properties are treated in a very specific manner. Garfinkel takes this as evidence that the natural sciences can be re-specified as sciences of practical action, because the presence of the qualifier "all" in the laws of physics must necessarily be accompanied with its accountability and observability, which depends on a set of practical actions that must be instructably observable to everyone.

Although the reproduction of this relatively easy experiment would seem rather straightforward—in fact, it is considered standard practice in high-school physics education—it turned out to be not so easy to re-constitute the reality of the phenomenon Galilei observed on the basis of his own instructions. Instead of interpreting this as a deficiency of those instructions, Garfinkel argues that this is a general property of the practical realization of any set of rules: the practical actions conducted in relation to a set of rules are themselves not captured within those, but rules function rather as norms that are oriented to in the flow of practical action (Ibid. 270). The practical circumstances within which experiments are conducted are—precisely because of the generalizability of the set of instructions—not contained within the rules that are oriented to in practical action. Accordingly, the experimental realization of the phenomena of physics is not a matter of rule-following but occurs in the interaction between the situatedness of practical actions in relation to the rules that are realized within this situatedness. The reality of the objects within scientific practice is, therefore, a continuous interaction between the indexical nature of the here and now and the potential generalizability of this here and now, realized by orienting to the rules designed to realize them.

Garfinkel's idea that the phenomena realized within scientific practices are constituted through the continuous interactions between generalizable rules and practical actions already points to the fact that science cannot be reduced to practical actions, since orienting toward rules requires an understanding of the relevant concepts that are contained within them. Furthermore, the rules that scientists oriented to are often *implicitly* presented rather than made *explicit* in concrete manuals, indicating that a shared epistemic stance about how to realize a specific scientific object is already present before a specific experiment is conducted. Integrating the hermeneutic approach of technological mediation into EM can also help reveal these aspects of scientific practice, thereby taking into account the theoretical-interpretative way in which scientific objects are constituted (i.e., the adjective "scientific" in "scientific practice").

The relevant concepts contained within the rules that are oriented to in scientific practices have become increasingly complex. As Lynch and Bogen have argued, in response to Garfinkel's program, and as continued by the conversation analyst Harvey Sacks, Galilean observation science should no longer be considered the model for studies of scientific practices (1994). As I have attempted to clarify in section 4.2, through a discussion of Bachelard's concept of *phenomenotechnique*, the increasing complexity of scientific objects is constituted because their realization fundamentally depends on how theoretical concepts can be made empirically present by technological means. This indicates that these concepts are not fixed but rather continuously change through how technological mediations are appropriated. In ethnomethodological terms, the theoretical concepts must be made indexically present in relation to relevant technologies. Making explicit the concepts present in the rules that are oriented to and how they are made interactionally present in relation to technological mediations require an understanding of the hermeneutics of how the "empirical" is interpreted in terms of existing theoretical concepts.

The constitution of scientific objects occurs through the continuous oscillation between existing theoretical concepts and how they are mediated by how they are made present within a phenomenotechnique. Accordingly, the appropriation of technological mediations can be made witnessably present by revealing the relevant concepts used to interpret the phenomena that are perceptually encountered within scientific practices and how their meaning becomes either contestable or more stable in this process. Doing so requires studying not only the practical actions that scientists undertake but also the theoretical-hermeneutical processes through which the reality that scientists study is constituted. In both cases, the constitution of this reality can be understood in terms of the norms that are oriented to in the making present of scientific objects within the relation between scientists and technologies.

6.5 CONCLUSION: A HERMENEUTIC APPROACH TO TECHNOLOGICAL MEDIATIONS

In this chapter, I have attempted to develop an approach to the empirical study of the appropriation of technological mediations in scientific practices. First, I introduce EM as an approach allowing to study the stabilization of reality as a continuous practical accomplishment. In scientific practice, the accomplishment of this shared reality largely occurs through how scientists hold each other accountable for epistemic judgments in conversation. However, the world in which these judgments occur is co-constituted by technologies that allow scientists to establish a relation with the phenomena they study. As such, an EM of scientific practice must be complemented with a postphenomenological approach capable of revealing how technologies mediate how the phenomena that scientists study become present.

Through a combination of postphenomenology and EM, I have suggested an empirical approach for investigating how scientific objects are constituted by establishing a relation with what Heidegger termed the "present-at-hand."

More technically, I have suggested that scientific practices must be understood as practices that explicitly aim to overcome the indexicality of its situatedness through the constitution of Galilean objects. Trivially, scientists cannot perceive their objects of interest without the presence of technologies. However, when "observation" is understood in terms of the constitution of a Galilean object rather than as a single act of perception, we must attempt to understand it as a process of *shaping* rather than simply an act of *seeing*, which complicates how technologies allow scientists to relate to the objects they study. The constitution of Galilean objects requires the continuous maintenance of shared norms that allow for transcending indexicality. These norms are made accountably present in the discursive interactions between scientists. Accordingly, when studying the conversations between scientists, the (often tacit) norms that are oriented to in the constitution of a shared reality can be made explicit.

For example, in relation to fMRI, specific norms come into being that constitute the reality of the scientific object "human cognition." Observing "cognition" is not merely a perceptual act facilitated by the presence of fMRI that presents it visually, but instead the product of an epistemic negotiation that encompasses the workings of the scanner, the reliability of the analysis used to transform measurements into images, and the establishment of an explicit link between the visual object (brain scan) and an interpretation of a specific cognitive function. In this case, the practical accomplishment of a shared reality—in this case human cognitive functioning—occurs through (i) the perceptual presentation of a visual object that is continuously (re-)appropriated by (ii) addressing the quality of its production and (iii) using a set of relevant concepts to make sense of its content. The structure of this accomplishment is not given but takes shape through how the perceptual phenomena encountered in scientific practices are interpreted. When understanding these aspects as technological mediations, it becomes possible to understand the norms that are oriented to in the constitution of human cognition in this context in terms of the hermeneutic process shaped by the technologies that scientists relate to.

Integrating EM into postphenomenology has allowed me to overcome some of the downsides of the philosophical perspectives discussed earlier in this book: (i) it overcomes the focus on individual perception often present in postphenomenology; (ii) it does not mystify a theoretical relation with the world—as was the case in Heidegger's phenomenology—because it makes researchable how technologies mediate how a relation with the present-at-hand is established; (iii) it allows to study how in relation with a specific phenomenotechnology, an epistemological rupture with everyday life is established without assuming—as Bachelard did—an already existing form of scientific rationality; and (iv) studying the norms to which scientists orient when overcoming the indexicality of scientific practice does not assume the presence of a specific scientific fact as is the case in Latour's semiotic approach. Because of this, an empirical investigation into how technological mediations are appropriated can be considered to open up a space for studying science-in-the-making.

Admittedly, my proposal that a combination of postphenomenology and EM can be used to study the appropriation of technological mediations has thus far been largely theoretical. However, the specificities of this approach are best seen when implemented. In the next two chapters, I do so by presenting two case studies in which the approach sketched in this chapter is applied. Both case studies focus on neuroscientific practices that are concerned with the study of human cognition. As will be demonstrated in the next two chapters, the different technologies used in different neuroscientific practices give rise to different epistemic norms, thereby each constituting the reality of "human cognition" in a different manner.

NOTES

1. Thoroughly reviewing the specificities of Garfinkel's empirical studies of the maintenance of a shared reality exceeds the purposes of this chapter. See, for example, Garfinkel (1967, 38–43, 79–88). For further discussion, see, for example, Heritage (1984, 89–97).

2. This already presupposes an interpretation of CA that slightly deviates from the traditional aim of CA to study conversations in real time by focusing on phenomena such as turn-taking and further general characters of the sequential organization of conversations. For a more detailed discussion about the relationship between Lynch's approach and CA, see, for example, Lynch (1985, 6–12), Ten Have (1990).

Chapter 7

Constituting "Visual Attention" in the Cognitive Neurosciences

INTRODUCTION: THE PROMISE OF THE "NEURO"

The promise of cognitive neuroscience is to explain human cognition in terms of its underlying neurophysiological interactions. A driving force in the development of this scientific domain is the idea that a unified theory of the human mind can be established with the help of the neurosciences. This is reflected in proposals for projects such as *Mindscope* (Koch et al. 2015) and *The Decade of the Mind* (Albus et al. 2007). Typically, it is stressed that the neurosciences have the potential to mitigate human suffering from mental and physical neurological diseases by developing "a mathematical and predictive framework of how all elements [of the brain] fit together and act as a whole to give rise to intelligence and consciousness" (Koch et al. 2015, 39).

Brosnan and Michael (2014) have argued that this promise of future merit binds different neuroscientific practices together and allows for the existence of "neuro" to endure. In this chapter, an important development within cognitive neuroscience that embodies such a promise is studied: Non-Invasive Brain Stimulation (NIBS). To do so, I followed a group of cognitive neuroscientists who are among the first to combine EEG with several forms of NIBS in the study of visual attention. The analysis focuses on how visual attention is constituted in the interaction between scientists who collectively relate to NIBS technologies through which human cognition is disclosed in terms of the human brain.

From its early use onward, the promise of NIBS is to establish *causal* relationships between brain processes and cognitive processes by stimulating the brain and observing the corresponding behavioral change (BC) (Sack 2006; Walsh and Cowey 2000). In this way, it is thought that neurophysiological processes that are functionally relevant to particular cognitive processes can be identified. However, NIBS does not allow researchers to *observe* neurophysiological processes and must be coupled with imaging technologies, such as fMRI or EEG for the visualization of neurophysiological change (NPC) (Herrmann et al. 2016; Miniussi et al. 2013).

In this chapter, I use the approach developed throughout this book to show that EEG, NIBS, and the combination of NIBS and EEG each give rise to a different constitution of visual attention, of which the reconcilability is not self-evident. The different technologies do not so much offer different perspectives on one singular object that can be identified as "visual attention" but rather make scientists relate to different provisional scientific objects that must be actively stabilized. As I show, the relation to these objects is mediated by the specific technologies used. I argue that the norms that scientists orient to when relating to the human brain differ per technology, which challenges the idea that the use of different technologies will eventually converge into one coherent picture of cognitive functions, such as visual attention.

The analysis shows that NIBS technologies make researchers strongly orient to a norm of causality that is weighed against a norm of "reality" (i.e., the extent to which experimental results can be extrapolated to situations outside of the laboratory). As a consequence, NIBS actively shapes how visual attention is constituted in neuroscientific practice. I subsequently demonstrate how the underlying interpretative framework of the cognitive neurosciences calls for the combination of NIBS and EEG and argue that the combination of NIBS and EEG mediates how neuroscientists orient themselves to the norms of causality and reality. The introduction of a new technology significantly shapes how the trade-off between causality and reality is managed, thereby constituting visual attention in a different manner. I conclude by suggesting that how researchers orient themselves to the norms of causality and reality during the constitution of visual attention (i) is mediated within the relations between scientists and technologies and that (ii) "causality" and "reality" are not explicit norms that are straightforwardly complied with but are implicitly oriented to when constituting visual attention as a neuroscientific phenomenon.

7.1 THE TECHNOLOGICALLY MEDIATED WAY IN WHICH COGNITIVE FUNCTIONS BECOME PRESENT IN SCIENTIFIC PRACTICE

Before starting to analyze this case-study, let me briefly recap some relevant concepts of the postphenomenological approach to scientific practice. In modern science, objects of research typically become visible to scientists in interactions with technological instruments. For instance, observing the human brain is impossible without imaging technologies such as MRI. Similarly, observing human cognition in neuroscientific terms depends on technologies such as fMRI, EEG, and NIBS. These technologies are central in translating relevant phenomena into candidates for human interpretation. In the words of the postphenomenologist Don Ihde, a "*translation* is a technological transformation of a phenomenon into a readable image (Emphases in original.). This is one analog to a hermeneutic process, except in this case it is a *material* hermeneutic process" (1991, 56). These translations can be understood as *technological mediations* that structure how reality becomes perceptible to scientists (Ihde 1991, 1998, 2009). Ihde suggests that, to understand modern scientific practice, philosophers must focus on these hermeneutic processes that give rise to the existence of scientific objects.

Although technologies help shape relations between scientists and the reality they study, they do not *determine* the objects of research. Rather, technologies *mediate* the reality that scientists study: in the relation between humans and technologies, a specific knower (subject) and thing known (object) are constituted, but the structure of the relation between the two is not fixed (Verbeek 2005, 130). Different technologies disclose the reality that scientists study differently and constitute different types of scientific observers. From this perspective, understanding how scientists relate to technologies becomes a way to understand how scientists and the objects they study mutually constitute one another. Ihde emphasizes that technologies make reality perceptually present in a specific way and understands technological mediations in scientific practice in terms of the constitution of a specific perceptual object that becomes visible for an embodied observer. This has inspired postphenomenological studies showing how images condition debates in the neurosciences (Rosenberger 2011) and how scientific observations of the surface of Mars are mediated by imaging technologies such as the Mars Orbiter Camera (Rosenberger 2008, 2013).

In emphasizing the actively mediating role of technologies, the postphenomenological approach has tended to overlook how human beings subsequently understand and act upon these mediations. Analyses of technological mediations deserve to be complemented with analyzing human appropriations *in actu* (Verbeek 2016, 191). The constitution of visual attention as an object of research depends not only on how brain activity becomes perceptually present through technological mediations but also on how groups of scientists collectively appropriate these mediations. As discussed in more detail below, I suggest that CA allows us to capture both the implicit and explicit ways in which technological mediations are appropriated *in actu* (Ibid. 195).[1] Thus far, the empirical study of how technological mediations are appropriated in scientific practices has not been a central concern in postphenomenology. In this chapter, I adopt the challenge of studying how technological mediations are appropriated by conducting an empirical study of how visual attention is constituted in neuroscientific practice.

Scientific Instruments in Neuroscientific Practice

The empirical study of laboratory practices has been a central interest of STS since its early days (e.g., Knorr-Cetina 1999; Knorr-Cetina and Mulkay 1983; Gilbert and Mulkay 1984; Latour and Woolgar 1986). Interest in neuroscientific practices increased after the designation of the Decade of the Brain (1990–1999), a large, publicly funded big science project devoted to the study of the human brain (e.g., Littlefield 2009; Littlefield and Johnson 2012; Pickersgill and Van Keulen 2012). Since the increased popularity of the neurosciences is closely related to technological developments that visualize brain processes in vivo, the role of brain imaging technologies in scientific practices has primarily been a subject of investigation (e.g., Beaulieu 2002). Critical studies have focused on how neuroscientists unjustifiably tend to reduce human beings and personalities to their underlying neurophysiology (e.g., Choudhury and Slaby 2012; Dumit 2004).

Furthermore, it has been shown that brain scans should not be understood as allowing to "peak into the human mind" but that understanding human cognition in terms of its underlying mechanisms is dependent on active processes of human interpretation (e.g., Roskies 2007). For example, it has been argued that the "seeing" of brain scans has been conceptualized as the product of organized embodied actions (Alač 2008), as the tacit knowledge needed to interpret brain scans (Gross 2009; Heiberg Engel 2008). While such analyses convincingly show that the interpretation of brain scans presupposes a specific social and embodied organization, they often surpass the question of how technologies help shape the *objects* of scientific research. In this chapter, I examine this latter question by investigating how visual attention is shaped in the relations between humans and technologies. I focus on how brain imaging technologies constitute the scientific object "visual attention" in a very specific way and shape how it becomes a reference point of collective orientation. From this perspective, scientific objects are never fixed entities but are rather *tentative* objects that are subject to continuous revision. In line with recent conversation analytic studies, we show that the coming into being of such objects can be understood as *practical accomplishments*. In other words, scientific objects such as visual attention "emerge and [are] established, perceived and understood, *from* processes and trajectories of social interaction" (Nevile et al. 2014, 17) (Emphasis in original). My analysis shows that the processes and trajectories of social interaction are mediated by the technologies that scientists use and, accordingly, mediate the epistemic norms scientists hold each other accountable for.

Conversation Analysis and Normative Accountability

I use CA to study the constitution of visual attention. Related to EM, CA (Sacks 1992; Lynch 1985, 1988; cf. Te Molder and Potter 2005) studies from a participant's perspective how people use language as a medium to construct a reality in which meaningful actions are performed. Like EM, CA is interested in the rules and norms that people use and hold each other accountable for. Both EM (Garfinkel 1967) and CA (Sacks 1992; Te Molder and Potter 2005) define everyday life as deeply and inherently normative.

According to conversation analysts, normativity is not so much the consequence of people obeying certain rules but is rather constituted by the fact that people *orient themselves toward* norms: actions are described in ways that display their status with regard to some rule or expectation (Heritage 1984). When categorizing events as departing from some normative order, people account for that deviation, thereby making routinely hidden norms available for inspection by researchers. By orienting themselves to these norms, participants organize their conversations around what they routinely interpret as normatively appropriate conduct (Ibid. 115–120). From this perspective, the practical accomplishment of an object is simultaneously a normative accomplishment. Building on this line of research, I analyze in this chapter how visual attention is constituted within the process in which scientists make themselves accountable to specific epistemic norms, which in turn take shape in the interaction with technologies.

CA allows to analyze interactions between scientists in real time, which enables to show the norms to which they orient in the active constitution of visual attention. Because these norms are constantly (re)negotiated, visual attention is interpreted not as a rigid entity but as an accomplishment that is subject to continuous change (Koschmann and Zemel 2014; Nevile et al. 2014). Consequently, scientific objects do not appear as stable referents but are dynamically constituted in the interactions between scientists and in the relations between scientists and technologies.

7.2 DATA AND METHODS

The data consists of video recordings of discussions between members of a group of researchers in the cognitive neurosciences during their weekly laboratory meetings. The material was recorded over a period of two months. I transcribed and examined the recordings with a primary focus on the verbal exchanges. A supplementary analysis of the video material was performed whenever it appeared necessary for a full understanding of what was said and done. The names of the participants have been altered, and the date and time indications have been deleted to minimize the possibility of identification. Institutional ethical approval was granted before the period of data collection. The data corpus consists of recordings of five laboratory meetings, comprising a total of approximately 500 minutes of recorded material.

The recorded meetings were transcribed Jeffersonian style[2] using ELAN.[3] The primary part of the data corpus was transcribed to a words-only level, but relevant parts were transcribed in more detail, including the pauses and overlap between speakers. The original conversations were in English, but most participants were non-native speakers. The conversations have not been translated into grammatically correct sentences, because this would undermine the objective of studying the constitution of objects in real-life conversations among participants.

In line with the principles of CA, my approach has been participant-centered (i.e., I examined how participants in the conversation actively constructed a meaningful reality) (Lynch 1993). In addition, I began with the assumption of technological mediation theory that considering technologies (such as NIBS or EEG) is crucial when it comes to understanding the constitution of objects. Accordingly, the analysis has focused on those elements in the conversations where the relation between the use of a particular technology and the constitution of visual attention was the subject of discussion.

7.3 NIBS, EEG, AND VISUAL ATTENTION

NIBS includes a set of different technologies that stimulate the brain at different levels. Neuroscientists use NIBS to study the neurophysiology underlying human cognition. In the present case, they aim to establish a causal relation between neurophysiological activity and our ability to detect visual stimuli. Visual attention is typically measured using an eye-tracking device. During experimental trials, participants are asked to fixate on a red cross that appears at different places on a screen. As experiments typically involve more than one participant, the relation between NIBS and visual attention is described in terms of the average of the statistical differences within a group of participants. Hence, observing visual attention requires the translation of the result of the stimulation into a graph that visualizes the average behavioral performance of the participants.

Specific to NIBS technologies is that they allow for the alteration of neurophysiological activity and that they are assumed to induce BC. Depending on the location and intensity of the stimulation, researchers expect that the participant's ability to fixate on the red cross will increase or decrease. The effect of NIBS is conceptualized in terms of the statistical difference between the participant's behavioral performance without stimulation and the behavioral performance during stimulation. Actual stimulation is often compared with a so-called sham condition, in which researchers pretend to stimulate the participant's brain. A comparison between the sham condition and the actual stimulation serves to filter out the BCs that are caused not by the induced stimulation but by the unusual circumstances in which experimental trials occur.

The neurophysiology underlying BC cannot be observed using NIBS. Observing NPCs requires one to use NIBS in combination with imaging technologies such as EEG and fMRI. Only in this way can researchers detect the neurophysiological patterns that correlate with the BCs occurring after stimulation. Successful integration of EEG and NIBS allows researchers to infer that the NPCs observed are actually caused by the NIBS technology, instead of being a random effect of other environmental stimuli.

7.4 CONSTITUTING VISUAL ATTENTION THROUGH NIBS

The causal reasoning instantiated by NIBS can be understood as follows: a stimulation technology (NIBS) causes BC (see figure 7.1).

By stimulating the brain at a specific location at a specific frequency, researchers can test hypotheses on how stimulation influences behavioral performance. For example, stimulating a specific brain area might increase or decrease the participant's ability to fixate on a red cross that is shown on a screen. In this way, visual attention becomes present as an object that can be manipulated by the researcher, such that a causal relationship between the stimulation and the BC can be established.

A crucial problem that the researchers face is how visual attention can appear as an object that can be causally reasoned about. This requires the careful construction of causality within a specific experimental setup. Researchers are very much aware of their own acts of construction, which makes them constantly question the reality of their own constructions. "Too much causality" is often interpreted as an indication of a "lack of reality." Thus, creating visual attention in terms of causality requires the management of potentially conflicting normative ideals that researchers hold each other accountable for: the norm of "causality" and the norm of "reality" (i.e., the extent to which experimental results can be extrapolated to the world beyond the laboratory). In the analysis, I show how this problem develops in relation to the technologies used and how it is managed in the interactions between scientists. In doing so, I highlight that the construction of visual attention is the product of a trade-off between normative conceptions of both causality

Figure 7.1 Causal Reasoning in NIBS Practice. *Source*: Figure created by author.

and reality. Contrary to a traditional take on science as a project that cumulatively approaches the truth about a fixed object in a stable reality, the reality of visual attention is continuously (re-)established in the negotiation between the norms of "causality" and "reality."

Consider the following fragment of a conversation between Mike, Pete, and Suzy that illustrates such negotiations. Suzy has just proposed a new experimental framework for studying visual attention that would be more in accordance with reality (i.e., more "ecologically valid" [line 28]). For that purpose, Suzy proposes the integration of "distractors" into the experimental design:

Excerpt 1:

Mike: m yeah but [if=
Mary: [yeah
Mike: =all it means adding so::me (.) random (0.5) pictures on both sides of the
visual field (.) and the (creations) still as they would were
Pete: I- I- I think that those paradigms are quite difficu::lt in terms of interpretation
(.) becau::se you never know whether you have an effect on the target
or the distractor that leads to certain behavioral change (.)
(16 lines omitted)
Pete: yea-yeah emmhe I mean distractors are interesting because the-they are
ecologically valid right? They are distractors over time an- an- an- sup-
pressing something that's not the::re is also a bit of a funny thing[right?=
Suzy: [yeah
Pete: so I completely get where you're coming from (0.5) it's just experimentally
speaki::ng adding this makes it a less well-controlled design

The proposal to integrate distractors is positively picked up upon by Mike in lines 3–4, who treats it as a minor adjustment to the existing experimental setup, namely, as only involving the addition of "some random pictures" (line 3). Pete, however, treats these distractors in an entirely different way when arguing that their inclusion would make the experimental setup "quite difficult in terms of interpretation" (lines 5–6), because the source of the BC can no longer be detected (lines 6–7). In other words, adding distractors complicates the "NIBS → BC" logic embodied in NIBS technologies. What Mike considers to be "some random pictures" is treated by Pete as seriously undermining the normative ideal of constructing visual attention in terms of causality. This requires a degree of precision that Pete argues to be incompatible with Suzy's proposed framework.

Pete agrees with Suzy that distractors would make the experiment "ecologically valid" (line 10) because in real life, people continuously suppress the presence of irrelevant yet potentially distractive stimuli (lines 10–11). However, adding distractions also makes the experimental setup less well-controlled (lines 13–14), which makes it no longer possible to interpret the experimental outcome in terms of the causal relation between NIBS and BC. The experimental setup must follow the logic instantiated by NIBS, which constitutes a norm of causality to which scientists orient in the construction of their object of research. Making the experiment more ecologically valid is not an option, because it comes at the expense of no longer being able to understand the experiment in causal terms. This implies that in the constitution of visual attention, neuroscientists hold each other normatively accountable for maintaining the causal structure of the experimental setup, and they do so at the expense of its "realness": the extent to which experimental results can be extrapolated to conditions outside of the laboratory.

The analysis thus suggests that the norm of causality competes with normative conceptions of the reality of a constructed object, such as its ecological validity. This indicates that when the constitution of visual attention is mediated by NIBS, (i) constituting visual attention can be seen as the interactional outcome of a trade-off between "reality" and "causality," that is, as two in practice conflicting norms to which researchers hold each other accountable; (ii) the norm of causality more strongly structures how visual attention is constituted than the norm of reality; and (iii) the norm of causality is not reflected upon (i.e., is not the subject of any explicit deliberation), and remains unproblematized.

7.5 COMBINING NIBS AND EEG: COMPLICATING CAUSALITY

NIBS technologies constitute visual attention in a specific manner because they embody the logic of a causal relationship between brain stimulation and BC (see figure 7.2.1).

When combining NIBS with EEG, the construction of visual attention becomes increasingly complex. Schematically, an element is added to the causal reasoning process that also accounts for NPC: NIBS causes NPC which in turn causes BC (see figure 7.2.2). To make the *observation* of NPC possible, the

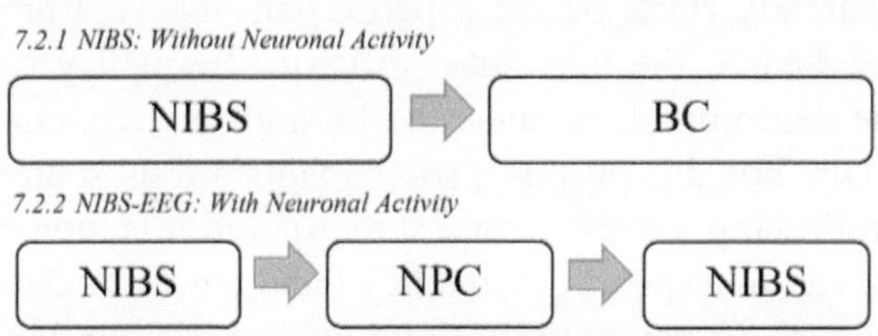

Figure 7.2 Causal Reasoning in NIBS-EEG Practices. *Source:* Figure created by author.

combination of NIBS with a brain imaging technology is required. In the present case, the researchers chose to combine NIBS with EEG. The combination of these technologies requires researchers to demonstrate a more complex causal relationship between brain stimulation and neurophysiological interactions.

In the group of researchers under study, EEG is combined with a specific form of NIBS, *transcranial alternate current stimulation* (TACS), which allows for the direct stimulation of oscillatory brain activity (Herrmann et al. 2016). Oscillations can be defined in terms of frequency bands and occur constantly without the presence of experimental stimuli. It is assumed that these different frequency bands correspond to specific cognitive processes (e.g., Ward 2003). The stimulation of a specific frequency band through TACS therefore opens the possibility of establishing a causal link between oscillations and specific cognitive processes.

Combining EEG and TACS constitutes a different causal logic, because now a *causal relation* between TACS, the NPC corresponding to the stimulation, and the corresponding BC must be established. The combination of EEG and TACS gives rise to the need to understand brain stimulation in terms of visualizations of the underlying neurophysiology, rather than in terms of the directly visible behavioral consequences. Also, a task stimulus can no longer be understood as only producing this or that neurophysiological response that can be measured by EEG, as it must understand this response precisely to the extent that it is caused by TACS. Not only should it become possible to observe the NPC occurring during the experiment, but the NPC must be interpreted as directly caused by TACS. As a consequence, causality can no longer be thought of as an interaction between two components but should be understood as distributed through an increasingly complex chain of different elements.

Constituting Visual Attention through TACS-EEG: Reality and Causality

The novel way in which visual attention is constituted through TACS-EEG becomes clearly visible when the researchers attempt to observe relevant neurophysiological activity and link it to the induced stimulation. This process of observation already starts before a participant receives brain stimulation. Before applying TACS, the relevant oscillatory frequency band of the participant must be determined, because the breadth of frequency bands varies among individuals. For this purpose, participants are presented with a cognitive task (say, following a red dot on a screen), and it is then analyzed, using EEG, which activity in a specific frequency band corresponds with the behavior of the participant. Consider the following conversation in which Pete and Suzy attempt to determine the breadth of a frequency band:

Excerpt 2:

Pete: I am seeing to your peak
Suzy: yeah and I mean it's like two hertz broad (.) they sa:y [hhh
Pete: [huh yeah yeah
Suzy: it it's two hertz but yeah [eh
Pete: [eh eh it's certainly not two hertz broa::d [right=
Suzy: [mmm
Pete: =because that's too nice a number to be true::
Suzy: yea yea yeah it's too nice=

In line 1, Pete reacts to a graph shown by Suzy, in which he identifies a peak that indicates the breadth of the frequency band of a participant. Suzy responds by specifying the exact breadth, when stating that it is "like 2 hertz broad" (line 2), corresponding to what "they say" (line 2). Because "2 hertz broad" is treated by Pete as an unreliable outcome (it is "too nice a number be true" [line 7]), I take it to refer to the average in breadth across participants in general. Pete's immediate rejection of the peak being 2-hertz broad shows that when determining individual frequency bands, researchers do not rely on external theoretical standards but consider that, in practice, measurements do not coincide with theoretical expectations. Contrary to the "beauty" of theoretical generalizations, individual electrophysiological activity is "messy." Accordingly, the norm of "reality" introduces an expectation that the experimental results should have a certain level of "messiness," which in turn functions as a reason to question the adequacy of statistical analyses. The criteria of "messiness" and "beauty" thus co-determine the (in) adequacy of experimental results and create certain normative expectations about the reality of the observed breadth of an individual frequency band.

Identifying an individual frequency band requires the *correlation* of neurophysiological activity and behavioral performance. When combining TACS and EEG, it must be ensured that oscillatory change can be understood as being *causally* produced by TACS, that is, not dependent on other variables. This requires moving beyond the correlational logic of EEG, which cannot rule out that a change in oscillatory activity is affected by something other than the induced stimulation. Furthermore, visual attention is no longer understood as the causal relation between TACS and BC but instead as the more complex causal relation between TACS, NPC, and BC (figure 7.2.2). As I show, the integration of NPC is not treated as a technical problem, but it spurs researchers to orient themselves to the norm of causality in a manner that narrows down how visual attention can be understood. In other words, the reality of visual attention is causally dependent on specific oscillatory patterns as shaped by the combination of TACS and EEG. Consider the following conversation between Mike, Mary, and Andrew:

Excerpt 3:

Mike: was there anything [that you guys have seen or not
Mary: [e::::hm no I don't think that it goes that fa:::r
(.) huhu so:: it was really .hh and I-I-I think even so what [colleague]
said and I see the same that even (.) between the different subjects for
one subject .hh the dorsal attention network is more linked toward
you kn- mmhmmh is more linked to five hertz and for the other its
more ten hertz [a::nd you know so th-the
Pete: [muh he he
Mary: [its really so we are with with the[the
Andrew: [is it just noise or is it more than noise I mean it
[could be a variable yeah
Mary: [yeah th-that's the e::h good question [.hhhh ha
Mike: [yea::h that's indeed also what I mean if we wanna push it
by scanning nine people it would be nice if it is on the basis of (0.5) some
let's say educated(.) guess that it works
(0.3)
Mary: yea::h
Mike: that <u>something</u> works huhu
Andrew: that TACS does anything (.) in the brain that would be nice

When discussing the outcome of the TACS-EEG experiment, Pete asks whether the effects of TACS can be "seen." Mary responds that she is not yet sure, because the analysis did not go "that far" (lines 2) (i.e., her analysis was unable to determine whether the oscillatory change could be causally linked to TACS). Too much difference exists between individual participants to determine whether the oscillatory change is caused by TACS (lines 4–7). This temporal understanding of seeing as a process over time instead of as an immediate act indicates that whether an object can be observed is dependent on how it is constituted. The norms of causality and reality are constitutive of both the act of seeing and the construction of the seen. This is reflected in Andrew's question of whether some type of systematics can be detected that accounts for the differences (lines 10–11). The seeing of visual attention is dependent on the extent to which it can be regarded as causally produced by TACS. Because no definite answer can be given to Andrew's questions, Mike states that at this point, he is unsure whether "something works" (line 18) (i.e., whether the individual neurophysiological differences can be causally explained in terms of the TACS stimulation).

The conversation ends with Mike displaying insecurity regarding whether TACS has stimulated the brain at all. This statement can only be meaningfully interpreted when the BC is to be understood in terms of not only TACS but also the underlying neurophysiology. Using the original NIBS logic, it would have been clear that the BC was caused by TACS. Using TACS-EEG, to establish a relation between TACS and BC is to understand the effect of TACS, but this understanding must be complemented with an account of the relationship between TACS and the neurophysiology. TACS-EEG research complicates the norm of causality, which must now be understood as the measurable causal relation between NIBS, NPC, and BC. Because of this change, it becomes possible to interpret the BC caused by TACS as not necessarily "doing anything in the brain" (line 19). The reality of visual attention becomes dependent on the causal norms that come into being in the relation between the researchers, and in the relation between the researchers and the technologies that they use.

With the introduction of EEG, the relation between TACS and BC becomes increasingly complex. This complexity sparks an uncertainty in TACS-EEG research, making it not straightforward and clear how NIBS, NPC, and BC are linked. Regardless of the complexity of the interactions between these three elements, researchers remain oriented to a norm of causality in TACS-EEG research. Consider the following interaction between Suzy and Pete:

Excerpt 4:

Pete: them and try to applic- apply them somehow hhh I am totally not
convinced [hahaha
Suzy: [no?
Pete: naaa nooo it- its its its a conflict of of research styles its its eeh (2.0)
huhuhuh
Suzy: tell me more about it?
Pete: no::: its its just I would always start with a proper EEG [experiment=
Suzy: [mhmm
Pete: =we have behavi[orally=
Suzy: [yea yea
Pete: =(0.3) things that can relate to physiological measures and and you
kno:::w and then you have (.) a correlation::: (0.3) of two things and
then you can (.)move on to causality with brain stimulation thats just
how I feel about it.

Pete indicates that he is uncertain about the approach taken in the laboratory when conducting TACS-EEG research (line 2). When identifying a "conflict of research styles" (lines 4–5), he contrasts the work in the laboratory with an approach that would "start with a proper EEG experiment" (line 7). Such an experiment would establish a correlation between behavior and physiology from which, he asserts, causality could be inferred; at least that is how Pete "feel[s] about it" (line 14). The use of phrases with a subjective connotation, such as "research styles" and "how I feel about it," indicates that his own approach does not provide an objective—for participants—manner to construct visual attention through the use of TACS-EEG. The constitution of visual attention in terms of TACS-EEG is dependent on the individual style of the researcher and is not treated as an "objective" solution that determines whether the NPC is caused by TACS—as would be the case if only NIBS were used to investigate the human brain. Although Pete suggests that the research should "start with a proper EEG experiment" (line 7)—indicating a clear starting point—he does not provide a justification for this idea other than to state that it would be the consequence of his preferred style of research.

The combination of intervening with NIBS, and imaging with EEG, requires the integration of the causal reasoning embodied in NIBS and the correlational reasoning embodied in EEG. In NIBS research, the construction of visual attention can best be understood as managing the trade-off between "causality" and "reality," in which causality is treated as unproblematically realized through NIBS. However, with the introduction of EEG, "seeing" visual attention is treated as problematic in this context, and the researchers cannot easily design a causal understanding of BCs in terms of the relation between the induced stimulation and NPC.

In summary, the constitution of visual attention, when investigated through TACS-EEG, reintroduces a norm of causality that appears as more complex than in the case of NIBS, because a new variable—NPC—must be integrated. Through this complexity, the norm of causality can no longer be unproblematically applied but is instead treated as relative to contingent norms, such as personal research styles that "allow to move from correlation to causality" (excerpt 4, lines 12–13). This suggests that how causality is made normatively relevant in the constitution of visual attention is mediated by the specific technologies involved. In turn, this leads to a different way through which the trade-off between causality and reality is managed in the constitution of visual attention.

7.6 CONCLUSION: TECHNOLOGICAL MEDIATIONS AND THE NORMATIVE EXPECTATION OF CAUSALITY

The analysis of how visual attention is constituted in cognitive neuroscientific research using NIBS, and combining NIBS and EEG, suggests a dominant orientation toward a cause-effect model of the relationship between brain and behavior and demonstrates how this orientation is taken for granted in the scientists' research routine. In the context of NIBS, I have argued that (i) constituting visual attention can be seen as the interactional outcome of a trade-off between "reality" and "causality"; (ii) the norm of "causality" more strongly structures how visual attention is constituted than the norm of "reality"; and (iii) the norm of causality remains unproblematized.

When NIBS is combined with EEG, the norm of "reality" suddenly becomes more complex, because reality is treated by the scientists as possessing a certain messiness that must be reflected in the observation of brain activity related to visual attention. I have demonstrated that the norm of causality can no longer be unproblematically applied in the context of NIBS-EEG. When brain activity is made visually available, researchers cannot straightforwardly assume a causal relationship between the induced stimulation and the observed BC. This, however, does not make researchers refrain from normatively orienting themselves to causality, but it rather spurs them to begin orienting themselves to contingent, subjective norms, such as individual research styles, thereby safeguarding the constitution of visual attention in causal terms (Gilbert and Mulkay 1984, 57). This suggests that (i) how researchers orient themselves to the norms of causality and reality during the constitution of visual attention is mediated within the relations between scientists and technologies and (ii) "causality" and "reality" function not as explicit norms but as implicit criteria, for which participants hold each other accountable, thus actively using these for the constitution of visual attention as a neuroscientific phenomenon.

Because visual attention is constituted differently in the case of combined NIBS and EEG than in the case of NIBS alone, it can be said that technologies are mediating how scientific objects come into being. Don Ihde (2009, 64) has argued that technologies significantly contribute to how scientists interpret the reality that they are studying and suggested understanding scientific practices as *material hermeneutic* practices. I have added to this work by revealing the interpretative processes through which technological mediations are appropriated (Verbeek 2016), thus showing *how* scientists translate a phenomenon made visible by material technologies into an *object of research*, namely, visual attention. What it means for an object of research to be causally affected is dependent upon the particular brain stimulation technique used.

Instead of understanding technologies as possessing preferred norms constituting preferred plans of action inscribed within them (e.g., Akrich 1992), my analysis does not assume that the norms that constitute the reality of scientific phenomena are inscribed in the technologies to which scientists relate. The study shows how "visual attention" is created in a trade-off between what is treated by the scientists themselves as the *conflicting* normative pair of "reality" and "causality." By studying *in actu* how visual attention is constituted within this specific normative orientation, I have not only shown *that* normative work is involved in scientific practice but also been able to trace *how* these norms shape how neuroscientists understand the objects that they study.

What are the consequences of this perspective for how the brain and mind are investigated in big science projects such as the Decade of the Brain and the Decade of the Mind? Can "a mathematical and predictive framework of how all elements [of the mind] fit together and act as a whole to give rise to intelligence and consciousness" (Koch et al. 2015, 39) indeed be developed? Critical approaches to the neurosciences have argued that neuroscientific research cannot be unproblematically used to solve societal problems (e.g., Dumit 2004). In their call for a Critical Neuroscience, Choudhury and Slaby advocate for critical engagement with neuroscientific research and argue that research on the impact and practices of neuroscientists should help to answer the question of "what kinds of ideas, hopes, methods, and institutions come together to produce what will count as facts about the brain" (2012, 4). By making explicit the role of the epistemic norms of causality and reality in the constitution of visual attention, I suggest that a specific normative framework constituting what counts as a cause-effect relationship comes into being as a consequence of how technological mediations are appropriated. Accordingly, such cause-effect relationships do not reflect an existing state of affairs external to the group of scientists but are normative accomplishments realized through how neuroscientists appropriate technological mediations.

Reflecting upon how scientists do this—that is, appropriating technological mediations—helps to reveal the interpretative—and normative—frameworks to which scientists orient themselves when constituting their objects of research. By revealing the material hermeneutic processes that structure the practices of the neuroscientists under study, I hope to have opened up a space for critical reflection on the reality that neuroscientists relate and the role of technologies in these processes. Because neuroscientists in the laboratory under study orient themselves to a linear norm of causality, other ways of understanding visual attention—and, more generally, other potentially valuable explanations of human behavior—are discarded, since these would not fit the desired causal model. Implicitly, this perspective rules out possible models in which the brain is not *governing* behavior and therefore becomes less central to what a human being is. Accordingly, the current popular understanding of the human brain as an autonomous agent that fully determines human behavior may not come as a surprise, as it actually reflects how neuroscientists themselves establish brain-behavior relationships in their scientific practices. By showing how the constitution of visual attention is dependent on a particular appropriation of technologies, such as NIBS and NIBS-EEG, I have shown that this understanding of human behavior is not necessarily constituted by orienting to an a priori concept of causality but is instead contingent on how a trade-off between the norms of "causality" and "reality" is managed in neuroscientific *practice*.

NOTES

1. Previous empirical postphenomenological studies on scientific imaging have nicely shown how technologies mediate perception in scientific practice in the context of microscopy in a cytology laboratory (Forss 2012) and in particle physics (Hasse 2008). In the present chapter, the focus is not primarily on scientific *imaging* and hence is less on how visualizations of scientific phenomena become perceptually available. Accordingly, the concept of "appropriation" is used precisely to understand how technologies mediate the constitution of scientific objects beyond the perceptual—although it does not at all aim to deny the fundamental importance of perceptual observation in scientific practices. To reveal the subtleties of how technological mediations are appropriated *in actu* requires to engage closely with the flow of action in which scientific practice occurs. This flow of action cannot be captured by using empirical methodologies such as interviewing techniques or participant observation (e.g., Aagaard 2015; Hasse 2008), but be fruitfully approached from the perspective of CA. For a more detailed discussion about postphenomenology and its relation to different empirical research methodologies, see Aagaard (2018).
2. See Appendix.
3. ELAN was developed by the Max Planck Institute for Psycholinguistics, The Language Archive, Nijmegen, The Netherlands (url: http://tla.mpi.nl/tools/tla-tools/elan/). For an explanation of ELAN, see Wittenburg et al. (2006).

Chapter 8

"Braining" Neuropsychiatric Experiments

INTRODUCTION: THE "NEURO" IN PSYCHIATRY

The widespread introduction of fMRI into psychology in the 1990s initiated a great optimism about the capability to finally unravel the mysteries of the human mind. Also, in psychiatry, brain imaging technologies were welcomed with great optimism as means to establish, with rigorous exactness, the cause(s) of patients' symptoms by allowing the tracing of mental disorders back to disorders in the functioning of the brain (Pickersgill 2011). In the DSM-IV, for example, it was suggested that the transformation from psychiatry into neuropsychiatry would finally grant psychiatry its desired "objectivity" by making obsolete the distinction between "mental" and "physical" disorders (APA 1994, xxi).

Initially, the use of brain imaging technologies in the neurosciences was guided by the assumption that specific cognitive functions could be mapped onto specific brain areas. From this perspective, the cause of a mental disorder could be found in a specific brain area that was not functioning "normally" (Vidal 2009). However, the current consensus in the neurosciences leans toward the idea that human cognition is realized within large networks of neurons spread across the brain (see, e.g., the *Human Connectome Project*), thereby generating complexity instead of the desired simplicity (e.g., Buckner et al. 2013; Menon 2011; Sporns 2014). As specific cognitive functions are thought of as being realized within networks of neurons, the ways in which mental disorders can be explained are also changing. Investigating the causes of mental disorders now requires researchers to identify the complex networks responsible for specific cognitive functions and discover patterns of "aberrant" organization within these networks.

Neuropsychiatric research searches to understand mental disorders in terms of underlying brain activity (e.g., Greicius 2008; Poldrack et al. 2012).

At the laboratory under study, the aim is to understand mental disorders, such as *attention deficit hyperactivity disorder* (ADHD), in terms of the neurophysiological interactions that constitute their symptoms. The promises of neuropsychiatry are (i) to offer an objective foundation grounding diagnostic processes and (ii) to allow for the prescription of clinical (pharmacological) treatments that specifically target the symptoms of a specific mental disorder. This model of understanding mental disorders posits neurophysiological activity as an explanatory cause of specific symptoms, thereby treating (the study of) the human brain as potentially *simplifying* the processes of psychiatric diagnosis and treatment (Insel and Cuthbert 2015).

In the previous chapter, I investigated how technological mediations are appropriated in the cognitive neurosciences and show how the reality of visual attention is constituted within the technologically mediated trade-off between the norms of causality and reality. In this chapter, I show, through an analysis of conversations between scientists who interpret brain scans, how the objects oriented to in neuropsychiatry are constituted through a different trade-off: one between complexity and simplicity. The question of how the *complexity* of the brain shapes neuropsychiatric practices must be answered relative to the aim of neuropsychiatry, namely, to develop simple explanatory models of mental disorders with a clear clinical purpose. I focus on how brain scans are interpreted in light of previous diagnostic categories and subsequently treated as incentives to develop experiments that are tailor-made to these categories. Through this focus on the appropriation of brain scans, I investigate how objects of neuropsychiatry are constituted within the presence of a specific phenomenotechnology.

This chapter departs from the idea that the use of fMRI in neuropsychiatry gives rise to a specific understanding of the brain *as* complex. How the human brain appears as complex in this context, I investigate through an analysis of conversations between a group of neuropsychiatrists in which they interpret experimental data. Data was obtained in a neuropsychiatry laboratory in a university hospital in the Netherlands and consists of approximately five hours of recorded conversations between neuropsychiatrists.[1]

The chapter is structured as follows: First, I suggest that how the human brain is understood as "complex" in neuropsychiatry is shaped through how scientists appropriate brain imaging technologies and that these appropriations can be studied from the perspective of EM. Secondly, I briefly discuss the relation between research in neuropsychiatry and the clinical function of psychiatry, focusing specifically on the relationship between the perceived complexity of mental disorders in terms of brain activity and the desired simplicity of clinical treatment. I subsequently demonstrate how the relationship between complexity and simplicity is actively managed by neuropsychiatrists, by showing how interpretations of experimental data are structured through an orientation to previous diagnostic labels. I then show how the appropriation of complexity becomes an incentive to develop experiments that are tailor-made to these already established diagnostic categories. To conclude, I suggest that ignoring the complexity introduced by network approaches to human cognition is a strategy to make relevant the work of neuropsychiatry for concrete individuals.

8.1 THE COMPLEXITY OF THE BRAIN AS TECHNOLOGICALLY MEDIATED

Neuropsychiatrists investigate the underlying neurophysiological substrates of mental disorders, and doing so requires the presence of brain imaging technologies such as EEG or fMRI that make it possible to observe brain activity in vivo. How neuropsychiatrists relate to (symptoms of) mental disorders is therefore dependent on how they relate to the technologies that they use to investigate those disorders. In the relation between scientists and technologies, specific interpretative categories and options for action and manipulation come into being that allow scientists to understand the object under study in a variety of ways (Ihde 1991, 137). Accordingly, the objects of scientific knowledge are constituted not only through the technological mediations that bring them within the domain of human perception but also through how scientists actively *appropriate* them within particular schemes of interpretation and action (Verbeek 2016).

Brain imaging technologies mediate how neuropsychiatrists understand mental disorders by making available in vivo the complex working of the human brain and its role in the development of mental disorders perceptually. However, the technologically mediated manner in which the complexity of the brain and how it helps realize specific forms of cognitive (dis)functioning does not occur in a vacuum but is relative to existing knowledge of mental disorders and brain functioning. Brain imaging technologies must be integrated into the existing goals and aims of neuropsychiatrists. In other words, when brain imaging technologies present the brain as a complex network, neuropsychiatrists are not simply confronted with a fait accompli. Rather, they must develop new interpretative strategies and plans of action against the background in which they understand mental disorders. Although the complexity of the brain cannot be simply surpassed, its role in relation to the understanding of mental disorders is not fixed. In the appropriation of the complexity of the brain, new ways appear to hypothesize how (symptoms of) mental disorders are realized and develop experimental circumstances in which such hypotheses can be tested.

The conception of the brain as a complex network is not new. Throughout history, complexity has often been singled out as a central feature of the human brain. For example, at the end of the nineteenth century, the famous Spanish histologist Santiago Ramón y Cajal developed an understanding of the brain of a complex organ on the basis of microscopic observations of the neural structure of brain cells. However, as the French epistemologist George Canguilhem has argued, philosophers interested in scientific knowledge should not be "thinking that persistent use of a particular term indicates an invariant underlying concept" (Canguilhem 1994, 32 op. cit. Rees 2016, 92). The question that should be posed, because of the situated nature of complexity, is how the use of fMRI in neuropsychiatry discloses the human brain *as* complex.

In the context of molecular biology, Hans-Jörg Rheinberger has argued that in the history of biology, crucial experiments typically developed parameters for simplification to be explanatory successful, while simultaneously retaining the complexity of the research object (e.g., genes) (1997). This implies a *trade-off* between simplicity and complexity in experimental research. Similarly, based on a critical reading of how pioneering synthetic biologists describe their work, Talia Dan-Cohen notes that researchers in synthetic biology are often trained as computer scientists and electrical engineers and deliberately ignore the biological complexity of life when developing models for designing and constructing novel organisms (2016). In this chapter, I show *how* a trade-off between simplicity and complexity is made in neuropsychiatric practice by exploring how neuropsychiatrists interpret experimental data. Because neuropsychiatric experiments and explanations make most of the aspects of the brain irrelevant, they do not unproblematically reflect the complexity of the brain. However, as becomes clear below, neuropsychiatrists do not so much *ignore* complexity as make it a *workable* element within the schemes through which they explain mental disorders. The chapter shows how the complexity of the brain is made, by laying bare how the specific technological mediations are appropriated by scientists on the ground.

An Ethnomethodological Perspective on the Objects of Neuropsychiatry

In chapter 6, I argued that studying appropriations of technological mediations requires an ethnomethodological perspective on the interaction between human beings and technologies. Rather than in isolation, technological mediations should be treated as belonging to processes within which scientific objects are constituted. In line with this perspective, I understand the coming into being of the objects of research within neuropsychiatry (i.e., mental disorders) as processes occurring within the relations between scientists and between scientists and technologies.

My analysis of the data is based on recordings of conversations between people working in the neuropsychiatry laboratory in which experimental data obtained by brain imaging technologies is interpreted. These conversations were transcribed using ELAN. Inspired by CA and EM, I use these real-time interactions to reveal the norms that are oriented to during these processes of interpretation (Garfinkel 1967; Heritage 1984).

In line with how Michael Lynch (1985) integrates CA into the study of scientific practices, I reveal how certain interpretative acts must be performed to maintain the continuity of the epistemic themes within scientific practices. Lynch argues that it is by orienting to projects addressing specific epistemic themes that the actions of scientific collectives are structured and within which certain features of the obtained data stand out as relevant. However, "the relatedness of actions in a project's sequence of actions [. . .] [is] not available in the continuous monitoring of the spectacle of events which where witnessed in the lab" (Lynch 1985, 55). Accordingly, "the continuity of a project is not a 'real-time' continuity [. . .] but the analysis [of practical activities] need to supply [. . .] a way of indexing successively displayed actions to their appropriate projects" (Ibid. 68). The level of detail in which the fragments are transcribed reflects my focus on analyzing how neuroscientists orient themselves to specific epistemic themes: I have chosen to transcribe overlaps between speakers and pauses to be able to analyze how orienting to an epistemic theme occurs *collectively*. I have refrained from a more detailed transcription level often used in CA, because my intention is not to analyze the structure of conversations proper but to instead reveal how epistemic themes are interactionally oriented to and made relevant.

When the project of neuropsychiatry can be understood as establishing (causal) relationships between brain activity and specific mental disorders, it is through this lens that the appropriation of technological mediations must be addressed. Accordingly, the central epistemic theme that I focus on in this chapter is how the perceived complexity of the brain is made to match the aim of neuropsychiatry to link patterns of brain functioning to symptoms of mental disorders.

8.2 NEUROPSYCHIATRY AND THE CLINICAL FUNCTION OF PSYCHIATRY: COMPLEXITY AND SIMPLICITY

Before specifically turning to how neuropsychiatrists appropriate technological mediations in practice, a brief explanation is necessary regarding the promise of neuropsychiatry in relation to existing psychiatric practices. As already indicated, the promise of integrating brain imaging technologies into neuropsychiatry is that they allow the measurement of brain activity, which will lead to more fine-grained diagnostic psychiatric processes in the long run. In the laboratory in which the data of this study was obtained, brain imaging technologies such as fMRI are primarily used to investigate how mental disorders such as *autism spectrum disorder* (ASD) or ADHD correlate with specific patterns of brain activity.

Diagnosing someone with ADHD, *obsessive-compulsive disorder* (OCD), or ASD is not to point to a singular entity but to categorize a multiplicity of different symptoms under a single header. For example, the DSM-V lists the following symptoms as diagnostic criteria for ASD: (i) persistent deficits in social communication and social interaction; (ii) restricted, repetitive patterns of behavior, interest, or activities; (iii) symptoms being present in early developmental period; (iv) symptoms causing clinically significant impairment; and (v) disturbances not being better explained by intellectual disability (APA 2013, 49–50). This indicates that neuropsychiatric research does not explicitly target mental disorders as a whole but works on the level of the symptoms associated with them. The laboratory in which the conversations were recorded has a strong focus on "restricted, repetitive patterns of behavior, interest, or activities," which is considered a symptom of either ADHD, OCD, or ASD. Accordingly, brain imaging technologies were primarily used to target aspects of the brain that were thought to be relevant for this specific type of behavior.

One would expect that fMRI, as well as the supposed objectivity associated with it, would be foundational when arriving at exact knowledge about the brain and its relation to certain pathologies. However, the previous psychiatric diagnosis, often based on the intuitive expert judgment of psychiatrists when meeting a patient for the first time, is leading, according to psychiatrists, when interpreting brain scans and the results of behavioral psychiatric tests (cf. Cohn 2012, 182–83). The supposed exactness of brain scans is thus rarely used to challenge previous diagnoses, but rather, the content, meaning, and relevance of brain scans in psychiatric research strongly depends on judgments made by means of the techniques of "traditional" (i.e., based on interviews and questionnaires) psychiatry.

While both of these schemes suggest a linearity in how brain scans are interpreted in (neuro-)psychiatry, in practice, the different ways of obtaining knowledge often intermingle, and interpretative processes follow a hermeneutic, rather than linear, pattern. Psychiatric tests are sometimes used to challenge brain scans; brain scans are sometimes used to challenge the expert judgments of psychiatrists; and earlier psychiatric judgments are sometimes used to challenge the authority of either psychiatric tests or brain scans. This interpretative pattern is guided by the question of whether the behavior of a patient, her brain visualized through fMRI, or certain results are indicative of (i) "normality" or "pathology" or (ii) a specific type of pathology (e.g., ASD, ADHD). It is only in this context that the supposedly exact knowledge of brain imaging technologies becomes relevant, and thereby necessarily intertwined, with ways to gather knowledge that can be deemed "less objective."

Building the Neurological Complexity of Mental Disorders

Mental disorders are cluster concepts that encompass a range of symptoms and make a wide range of different types of behavior fit into a single diagnostic category. This points to a tension in how brain imaging technologies are treated: on the one hand, a consensus exists among researchers that the brain and how it realizes specific cognitive functions must be understood in terms of neural networks that cannot be traced back to a specific origin and should not be associated with specific brain regions. This makes it increasingly difficult to understand (symptoms of) mental disorders in terms of their underlying neurology (cf. Fitzgerald

2014; Maung 2016). This complexity is treated as objectively displaying that the complexity of the human brain is a fact of nature. On the other hand, however, at the same time as being treated as an inevitable feature of human nature, the displayed complexity is treated as a manageable and malleable phenomenon that can be simplified to conduct experiments in neuropsychiatry. The following fragment is indicative of this tension:

Excerpt 1:[2,3]

Sally: it has been very helpful for my biological background of the behavior
that [eh we see
Mike: [it's not so straightforward [eh=
Sally: [no:::
Mike: [=and that again and that's what makes doing
research e:::h terribly complex
(3 lines omitted)
Sally: yes and that makes that you cannot say the putamen is my region of
interest and because it is different we understand how autism works e::h
because it is so extremely interwoven
Mike: yes yes yes I mean these loops remain [important=
Sally: [yes
Mike: [=of course and it might be possible
that one loop gets in one way or another more attention than another loop that's
possible regardless of whether they are more integrated into another system
it can still be the case that one gets suppressed and that has to be (0.5) otherwise
the behavior can <u>never</u> be explained
(3 lines omitted)
Sally: yes but it gives me now primarily the feeling that e::h you get the tendency
to pull in more and more and that eventually:: you end with yes the
brain is a large network and everything is [connected=
Mike: [hm ↑ hm
Sally: [=you need everything du::::h [that's
Mike: [yeah
Sally: [of course
true but it's e::h incredibly difficult to e::h consider regions in isolation
Mike: t- that's simply the case

In this conversation between a postdoctoral researcher (Sally) and a senior researcher and vice-director of the laboratory (Mike), the brain is being built as a complex organ that evades simple explanatory models of mental disorders. Sally was asked to write a review of how the neural networks associated with repetitive and stereotyped behavior map onto the anatomical regions of the brain. She states that doing research for this review was a "good way to see the biological background of the observed behavior" (lines 1–2). This is treated by Mike as indicating that the origins of behavior in the brain are not "so straightforward" (line 3), which makes doing research "terribly complex" (lines 5–6). Sally responds affirmatively to Mike and treats the complexity of the brain as preventing researchers from focusing on its specific regions (the putamen, line 8) and using deviances in specific regions as explanatory for autistic behavior (lines 9–10). Instead, she orients herself to the complexity of the brain as indicative of the fact that autism is "extremely interwoven" (line 10). Accordingly, complexity is oriented to as a norm that forces moving beyond the mapping of brain functions to specific brain regions.

In response to Sally, Mike suggests that the loops of activity observed in the brain may have an important explanatory function (line 11). "Loops" are patterns of brain activity moving across different areas of the brain, and specific patterns of loops are thought to be indicative of specific cognitive functions (e.g., Amso and Scerif 2015). Mike states that thinking in terms of loops still possesses some explanatory potential, because it allows the explanation of autistic behavior as constituted by one loop suppressing another (lines 16–17). If not, he says, the behavior observed in autism "can never be explained" (lines 17–18). Sally responds to this suggestion by stating that such an approach only leads to the "tendency to invoke more" and eventually leads to a view that "the brain is a large network and everything is connected" (line 22). Her consecutive "you need everything du::::h" (lines 24–25) displays the self-evidence of such network complexity yet clearly demonstrates annoyance concerning the lack of explanatory potential of such a perspective. In this case, the desired simplicity of using the brain as an explanation of autistic behavior is oriented to as a norm conflicting with the complexity of the brain. The conversation ends by Mike saying that this complexity "is simply the case" (line 29), indicating that there is no theoretical path out of this problem.

This conversation shows that neuropsychiatrists must manage a trade-off between two theoretically conflicting norms to integrate the "neuro" into psychiatry: on the one hand, they orient themselves to the norm of (neuro-) biological complexity, while on the other hand, they orient themselves to the desired simplicity of causal explanations of how symptoms of mental disorders are realized. This indicates that the complexity of the brain is not uncritically accepted but is appropriated and integrated into the existing aims of neuropsychiatry. In the next two sections, I show how this trade-off is managed by focusing on two interrelated epistemic concerns of the neuropsychiatrists under study: (i) on how brain scans are interpreted in terms of previous diagnoses and (ii) on how the brain scans are treated as an incentive to develop new ways of experimentation. This analysis reveals the sense in which the possibility of observing brain activity in vivo is made relevant for the purposes of neuropsychiatry.

8.3 THE ROLE OF DIAGNOSTIC LABELS IN THE INTERPRETATION OF EXPERIMENTAL DATA

The complexity of the human brain and its role in the realization of cognitive functioning is not only a matter of theoretical concern, as shown in Excerpt 1, but it also affects the sense in which brain scans are treated as objective. In this section, I discuss how researchers orient themselves to brain activity visualized through fMRI by analyzing two excerpts from a conversation in which the data of an experiment that aims to target "repetitive and stereotyped behavior" in individuals diagnosed with ADHD, OCD, or ASD is discussed.

At the laboratory under study, the perceived complexity of the brain mediates what is understood by "repetitive and stereotyped behavior" as a symptom of ADHD, OCD, or ASD. The symptom "repetitive and stereotyped behavior" obviously encompasses a wide range of heterogeneous instances of behavior. For example, in the DSM-V, examples of this symptom in the context of ASD range from "extreme distress at small changes" to "greeting rituals" (APA 2013, 50). However, these different types of behavior are treated as homogeneous, in the sense that they are different instances of repetitive and/or stereotyped behavior. This type of behavior is a symptom of ADHD, OCD, and ASD; hence, neuropsychiatrists often stipulate that, in one way or another, a continuum exists between different mental disorders in terms of this symptom (e.g., Hyman 2010; Lichtenstein et al. 2010). Yet, at the same time, the different diagnostic categories constitute a normative expectation that different patterns of brain activity are underlying the behavior in the different patient groups. Researchers thus must orient themselves to the complexity of the human brain such that it becomes relevant in terms of the simplicity of existing diagnostic categories. How the complexity of the brain is oriented to in this context, as well as the extent to which it is oriented to as objectively, becomes clear from my analysis of two excerpts in which brain scans are interpreted.

The first of these excerpts is taken from a conversation between the head of the laboratory (Mary), the researcher also encountered in excerpt 1 (Sally), and a PhD candidate under the direct supervision of Sally (John). The conversation focuses on how to interpret the data that John obtained from a recent experiment on impulsivity. The experiment used a "GO-NOGO" paradigm, individuals were asked to react as quickly as possible when presented with a stimulus, by either pushing on a green button (GO) or on a red button (NOGO). In doing so, the researchers aimed to study the extent to which presenting deviant stimuli influences a participant's ability to continue the task as instructed. It was hypothesized that individuals diagnosed with OCD, ASD, or ADHD have increased reaction times or make mistakes when presented with a deviant stimulus.

Excerpt 2:

Mary: I estimate that this will simply be (.) in terms of diagnostic differences
in brain activation in this go no go task eh between these patient groups
simply a negative study
Sally: yes
Mary: at this point we can no longer deny [this
John: [no
(9 lines omitted)
Mary: I would still like to know how these two groups are divided among the
three diagnostic groups ADHD ASD and OCD and is this something
John: yes this is the first part of the ROI analysis in which I e:::h actually these
three and there are some more in the other groups that I tested but e::h
in the first contrast nothing come out but I want to do it on contrast three
because on a whole brain level [something
Mary: [that something was there
(4.0)
Mary: that sounds like a good idea
(2.0)
Mary: but thus far (.) the eventual conclusion is thus [primarily
John: [yes haha that one
Sally: stupidly this would fit in into my impression of the literature at this point
and then it is really a pity that eh yes it is not the nicest things to write a
negative [study=
Mary: [no but is not unimportant
Sally: [=but it is not a real surprise that you do not find anything
in this group with this task a:::nd maybe it makes sort of sense to
(.) mention that ↑ just very explicitly

The excerpt begins with Mary stating that a conducted experiment does not display differences in brain activation patterns across the different patient groups (lines 1–2), which makes the experiment "a negative study" (line 3). Both Sally (line 4) and John (line 6) agree with this assessment. The adjective "negative" indicates that earlier psychiatric diagnostics function as norms that the conversational participants orient themselves to when developing experimental hypotheses. This constitutes an expectation that when experimental participants demonstrate a similar task performance, the underlying patterns of brain activity would differ between different patient groups.

Later, however, in response to Mary, John says that an option may still exist where differences are visible on a whole-brain level (line 13). Again, this assumption is guided by previous diagnostic labels, because it suggests that although the analysis focusing on a particular region of interest in the brain might not display differences in brain activation, these differences may be present on the level of the brain as a whole. While Mary responds to this by saying that "that sounds like a good idea" (line 16), she continues by restating her earlier conclusion that no difference across the patient groups exists (line 18), which John affirms (line 19). Even though the complex interactions on a whole-brain level might display differences across patient groups, this is not treated as potentially challenging the earlier conclusion. Hence, the complexity of brain activity on a whole-brain level is displaced as a reliable source for detecting differences between patient groups.

In the final part of the discussion, Sally states that this conclusion "stupidly fits her impression of the literature" (lines 20–21) and that while it is a pity to have to write a negative study (lines 21–22), the outcome is not "a real surprise" (line 24). Sally states that the significance of this study is to be found in the fact that no differences in brain activity can be found while using the "particular task" they have used and that it is worthwhile to "mention that very explicitly" (lines 27–28). Hence, on the one hand, the observed brain activity is treated as objectively showing that no differences can be identified across patient groups, yet on the other hand, this is not necessarily treated as a feature of the participant's brain activity as such but is made relative to the experimental circumstances in which brain activity was measured. The previous diagnosis of a specific mental disorder is thus invoked as a norm for interpreting the experimental data, as the measured brain activity is not assumed to objectively indicate a similarity across mental disorders but instead prompts a discussion about whether specific experimental circumstances can still "pull out" differences in brain activity. Accordingly, the complexity of the human brain must be negotiated in relation to the expectations constituted in relation with the type of mental disorder that is attributed to the brain under discussion.

This negotiation occurs in excerpt 3, where Sally and Marry continue to discuss the conceptual possibilities that might help to experimentally reveal the circumstances under which differences in brain activity can be detected:

Excerpt 3:

Sally: e::h what what the data from America would say is that almost the only
way to e::h get this type of behavior using this task is to put something
that is soo:: incredibly [salient=
Mary: [salient yes yes
Sally: =that it immediately sort of disturbs all behavior (.) yes and that that then
they can do task slightly worse and then you speak about really slightly
(6 lines omitted)
Mary: .hhhhhh yes and and and then the salience does something and you get
some movement and and then still there are no enormous differences as
far as I can see
(7 lines omitted)
Sally: you can e:h create some little movement for example by showing social
stimuli but then it might still be rather the [limbic system that is e:::h=
Mary: [that something's wrong with that
Sally: [=causing a very litte disbalance e:h in
the cognitive control system then that the cognitive control system itself is not properly
functioning
Mary: but how does this e::h fit when we look at the behavioral [data=
Sally: [yes
Mary: [=all the data from questionnaires
suggests that oh it seems that [actually=
Sally: [.hhhh yes:::
Mary: the three mental disorders are not really [discrete=
Sally: [no
Mary: [=but that there is a continuum but then we do these
sort of studies and take an adhd-task and we push and pull and we actually=
(all laugh)
Mary: get no movement in the ASD group that was supposed to be in the same
continuum and that is just:::

Sally suggests that on the basis of earlier obtained data, it might be helpful to include stimuli that are "incredibly salient" (line 3) to show behavioral differences across patient groups in this particular experimental design (lines 1–2). Earlier experiments showed that in terms of task performance, groups diagnosed with a specific (yet until now unspecified) mental disorder had a slightly worse task performance, but Sally admits that this difference was quite minor (line 6). In response, Mary states that integrating more salient stimuli causes a "little movement" but still does not display "enormous differences" between distinct mental disorders (lines 9–10). Again, previous diagnoses have a strong normative input: the expectation of the outcome of the experiment is that there are "enormous differences" that are "really clear."

In response, Sally suggests that presenting "more social stimuli" might cause some more "movement," because these might allow for investigating whether something is wrong in the "limbic system" of the individuals diagnosed with a mental disorder (line 13). She says that these different reactions to different types of stimuli are treated as new ways to understand the underlying causes of deviant behavior. They namely point to the possibility that this limbic system "brings the cognitive control system out of balance" rather than that something is wrong in the "cognitive control system itself" (lines 15–17). This indicates that different experimental circumstances reflect different hypotheses about how certain types of behavior are realized in the brain. Instead of that Sally orients herself to brain activity to find the root of deviant behavior, brain activity is treated as a malleable phenomenon reflecting the specificity of certain experimental circumstances.

After Sally has proposed a difference in terms of brain activity, Mary returns to the question of how the experimental data relates to the behavioral data obtained via questionnaires (lines 20–21). Mary treats the questionnaire data as relevant because it suggests that a continuum exists between the three mental disorders that are clinically manifest (lines 25–27). However, as she points out, when conducting their studies and using an "ADHD-task" (line 26), there is no movement in the ASD group, even after they "push and pull" (line 27) the data. She suggests that this is peculiar, because on the basis of the questionnaires, ASD and ADHD would be in "the same continuum" (line 29–30). This is where the discussion ends, and, as suggested by the laughter displayed in line 28, all of the participants treat this conclusion as rather uncomfortable. Again, rather than treating the experimental outcome and the observed brain activity as reflecting "facts of nature," it is oriented relative to earlier clinical diagnoses.

My analysis shows that diagnostic labels have a strong normative function in processes of data interpretation. Researchers actively attempt to reconcile brain activity with existing diagnostic labels, even when this seems difficult. As such, the visualized brain activity is attributed an objective status to the extent that it can be aligned with earlier diagnoses (cf. Cohn 2012). The absence of relevant differences in brain activity in different diagnostic groups is not treated as an "objective" fact but rather functions as an initiative for researchers to develop new experiments that allow for aligning the observed brain activity with previous diagnostic labels. In the next section, I explore how the relation between these two aspects is managed in the development of new experiments.

8.4 "BRAINING" PSYCHIATRIC EXPERIMENTS

In the previous section, I showed that researchers orient themselves to the complexity of the human brain in terms of previous diagnostics processes that have already classified the participants in the experiments as having a certain mental disorder. Only in light of previous diagnoses does it become possible to interpret the observed brain activity. In other words, the researchers must develop experiments that can tie neurophysiological activity to specific symptoms of mental disorders. As this section demonstrates, researchers develop a practice of "braining" psychiatric experiments to do so (i.e., they develop experimental setups in which brain imaging technologies are appropriated in such a way that they can be understood in terms of earlier psychiatric diagnoses). In doing so, the complex structure through which cognitive functions are realized is neglected and fMRI is treated as allowing the linkage of symptoms of mental disorders—in this case, impulsivity—to specific brain regions. Orienting to fMRI in this way assumes that human cognitive functions originate from particular brain areas and that fMRI allows the objective localization of specific cognitive acts in the brain in vivo.

In the excerpts analyzed in this section, how "repetitive and stereotyped behavior" can be experimentally investigated at the laboratory by linking it to the impulsive reactions of individuals when being confronted with something that is of their central interest is discussed. In a presentation on the experimental study of repetitive patterns of behavior in children diagnosed with ASD, a researcher at the laboratory explains why she makes this specific link along the following lines:

Excerpt 4:

Mary: so e::h everyone has an interest or a hobby could be sportcars
(5 lines omitted)
but in any case usually you should be able to function around that interest
so it should not impairing your daily life you should be able to function
at a job without being busy with sport cars all the time e::h but this is
different in autism
(4 lines omitted)
and that interest e::h may be so intense that it does impair them in their
daily lives so:: e::h
(5 lines omitted)
they are rather all consuming and they do limit social interactions when
they interfere with daily activities
(3 lines omitted)
so: e:h what we wanted to know is the behavioral and neural pattern of
impulse control to preferred interest first in healthy adults that is a pilot
study we did and then we wanted to know whether children with autism
would demonstrate a bias towards their restricted interests compared to
typically developing control children

Repetitive and stereotyped behavior is, in this experimental context, understood in terms of the specific interest of an individual with autism that "may be so intense that it does impair them in their daily lives" (lines 8–9). To a certain extent, the relation between individuals with autism and their interest is treated as being continuous with the hobbies and interests of healthy individuals (line 1). However, the behavior that individuals diagnosed with autism display in relation to those interests is treated as marking a difference, because they "are rather all consuming [. . .] and interfere with daily activities" (lines 11–12). This behavioral difference is oriented to as a norm that constitutes a break in the continuum between "healthy" individuals and individuals diagnosed with autism: children with autism "demonstrate a bias towards their restricted interests" (lines 17–18), which makes them "deviate from the behavioral and neural pattern of impulse control [. . .] in healthy adults" (lines 15–16). What is treated as the most relevant question in this context is how this hypothesis can be tested in terms of both the behavioral and neural differences between healthy individuals and individuals diagnosed with ASD. Hence, the difference between "healthy individuals" and "individuals with ASD" is already assumed by orienting to how individuals have been classified through earlier diagnostic processes.

Because how scientific experiments are materialized is mediated by fMRI, the difference between healthy individuals and individuals who are diagnosed with ASD must be reflected on a neurological level. Given the complexity of neuronal interactions and their relation to cognitive and emotional functioning, experiments must be designed to specifically target the neurological underpinnings of the "impairment" in the brain that constitute the deviant behavior in individuals diagnosed with ASD. Accordingly, the complexity of the human brain is made relative to specific interests of individuals and the associated impulsivity that is the consequence of them. Only when this situation is realized can the brain have an explanatory function within this type of neuropsychiatric research. Ironically, this requires researchers to align the perceived "objectivity" of the human brain as revealed through fMRI with the "subjective" interests of individuals to become relevant *as* objective knowledge. Accordingly, "braining" psychiatric experiments simultaneously requires the researchers to personalize psychiatric experiments.

As explained during the presentation, because the impairments that are observable in individuals with ASD are relative to the behavior of "healthy people," it is first necessary to find a way to study impulsivity in this latter group. Prior to the experiment, individuals were asked to choose from twenty-two different potential interests (ranging from trains to SpongeBob SquarePants) the three interests most to their liking. Subsequently, these "healthy" individuals performed a GO-NOGO task and were asked to refrain from pushing a button when confronted with one of the interests of their choosing. As a control condition, participants were asked to perform the same task but were instead confronted with different facial expressions and colors. Behaviorally speaking, these healthy individuals were equally good in performing this task under the three different conditions. However, the researcher argues that in terms of brain activity, a difference can be detected. Excerpt 5 demonstrates how the objectivity of fMRI is oriented to when substantiating this claim:

Excerpt 5:

Mary: e:::::h it was an fmri study so:::
(4 lines omitted)
basically the take away message here is that e:h areas were activated that
you would expect during this type of task so the visual cortex because
they are looking at things the motor cortex because they're pressing buttons
e:h and then e::h insula because you are looking at things that are
salient to you or at least eh presu e:h presumably salient to you e::h and
there was some interior frontal gyrus cognitive control activation
(11 lines omitted)
which was somewhat makes sense because probably that what is your
hobby is salient to you and e:h the insula e:h e:h is activated when seeing
things that are salient to you e::h which is why e::h e:h left insula activity
was higher but even though e:h there was a slight increase in brain activation
in healthy individuals are just able to inhibit themselves regardless

Interestingly, the fMRI results are treated as relevant not because they display something new and/or exciting but because they show that the "areas [of the brain] were activated that you would expect during this type of task" (lines 3–4), which is further specified in lines 3–8. Only the potential reason for the "interior frontal gyrus cognitive control activation" (line 8) is explicitly explained, indicating the presumed relevance of this brain area for this specific task. Activity in the insula makes sense because it "is activated when seeing things that are salient to you" (lines 11–12) yet is not constitutive of deviant behavior because "healthy individuals are just able to inhibit themselves regardless" (lines 15–16). Instead of providing new information on how behavior is neurologically constituted, fMRI is used in this experiment to validate whether the stimuli used in the experiment trigger activity in the brain areas they are supposed to trigger. fMRI is thus treated as offering objective certainty in the sense that it allows the validation of whether the experimental stimuli can be mapped onto brain regions associated with certain cognitive functions (e.g., looking is linked to the visual cortex, etc.). Thus, rather than being invoked as introducing complexity, fMRI is oriented to as allowing the causal linkage of specific brain areas to specific cognitive functions.

A similar function is ascribed to fMRI when the researcher discusses how the supposed "impairments" of individuals diagnosed with ASD can be made present experimentally:

Excerpt 6:

Mary: e::h these sort of relatively neutral stimuli or stimuli that are just not
of interest to an individual with autism e:h do not capture that change
in impulse control e:::h what does capture in individuals with autism's
interest is <u>what they</u> like
(3 lines omitted)
and two recent studies has also shown e::h that again e:h individuals with
autism show increased activity in left insula when viewing e:h images of
their special interests e:h and they also show increased activation in the
fusiform area which is an area that is traditionally been associated with
face recognition or f e::h and it it indicates some form of high visual
expertise basically
(5 lines omitted)
somehow they have great visual expertise in watching what is very interesting
to them thereby activating those regions that for us are activated
with faces and for them activated when watching what they like

Sally identifies as a potential problem that "relatively neutral stimuli [. . .] do not capture the change in impulse control" (lines 1–3). Rather, experimental stimuli should reflect "<u>what they</u> like" (line 4) (i.e., target impulsivity in relation to the specific restricted interests of the participants). She refers to other studies that showed evidence that "increased activity in left insula" was shown "when viewing images of their special interests" (lines 7–8). This resonates with the activation in healthy individuals when confronted with images of their interest. However, in addition, they display "increased activation in the fusiform area," which is traditionally associated with face recognition, and indicates high visual expertise (because healthy individuals are relatively good in recognizing faces). Accordingly, a parallel is drawn between the visual expertise that healthy individuals display in face recognition and the visual expertise of individuals with ASD "when watching what they like" (line 15). This suggests that *if* a GO-NOGO paradigm were used for investigation, the stimuli used in it should at least trigger brain activity in the "left insula" and "fusiform area." So, by linking specific cognitive functions to specific brain regions, fMRI is treated as introducing a norm to which experimental stimuli must conform: only when activity is present in the identified brain areas is the development of experimental stimuli treated as successful.

fMRI thus introduces a normative framework against which experimental stimuli are evaluated, eventually giving rise to the idea that these stimuli must be "tailor-made" relative to the subjective personal interests of individuals. Only in this way can prior distinctions between "healthy" individuals and individuals diagnosed with a mental disorder be maintained. The human brain is not oriented to in terms of its complexity in this context (as we would, for example, expect on the basis of excerpt 1). Instead, fMRI is treated as an incentive to develop experiments that are tailor-made to (confirm) previously established diagnostic categories.

8.5 CONCLUSION: THE OBJECTIVITY OF FMRI IN THE CONTEXT OF THE TRADE-OFF BETWEEN COMPLEXITY AND SIMPLICITY

By attempting to understand the perceived complexity of the human brain in terms of how mediating technologies such as fMRI are appropriated in neuropsychiatric practice, I have attempted to show how the possibility to visualize brain activity in vivo mediates how the brain is understood *as* complex. This has revealed, firstly, that visualizations of the brain are not treated as offering a straightforward form of objectivity, or as finally allowing the unraveling of the mysteries of the human psyche and the origins of mental disorders. Nor was fMRI treated as unproblematically allowing the interpretation of the human brain in terms of a complex network. A sharp contrast thus exists between the consensus that brain imaging technologies reveal the enormous complexity of the human brain and the actual practice of neuropsychiatry: complexity is managed to suit the pragmatic goals and aims at stake. This chapter highlighted two ways in which this can be done. First, diagnostic labels are oriented to as norms that structure how visualizations of brain activity are (to be) interpreted. A prior difference between "healthy" individuals and individuals diagnosed with certain mental disorders is thus presupposed for brain scans to become interpretable. Secondly, this prior difference informs how neuropsychiatric experiments are "brained": fMRI is treated an incentive to develop scientific experiments that are specifically tailored to confirm earlier established diagnostic differences.

The relevance of observing brain activity for the purposes of psychiatry, as well as how the fMRI is treated as offering objective knowledge, must be considered relative to how the trade-off between complexity and simplicity is managed in practice. As such, fMRI should not be considered a technique that allows psychiatrists to perceive brain activity but rather as belonging to the phenomenotechnology within which the objects of neuropsychiatry (i.e., [symptoms of] mental disorders) are constituted. While theoretically, the brain is oriented to as a complex organ (see excerpt 1), in their daily routines, researchers manage this complexity vis-à-vis the pragmatic aims of (i) distinguishing between the "normal" and the "pathological" and (ii) making this difference neurologically present in scientific experiments. Inevitably, a trade-off between the organic complexity of the human brain and the desired simplicity of explanations of (symptoms of) mental disorders must be managed. Accordingly, the technological mediation of scientific knowledge should be understood in terms of how mediating technologies function against an existing background of pragmatic aims and existing theoretical knowledge. Acknowledging this helps to clarify the problematics that neuropsychiatry faces when the brain is to be understood as a complex network. The analysis shows that the complexity of the brain must be understood not in isolation but rather in terms of how it is managed and made relevant in the daily routines of neuropsychiatrists.

Currently, neuroscientists are confronted with the ideal that unraveling the complexity of the brain will help mitigate the sufferings of people with mental disorders. However, at the same time, whether this ideal has a biological or cultural origin is unclear. For example, Cornelius Borck has shown that "the complexity so easily assumed to apply to the brain is very much an attributed quality, reflecting a cultural expectation and a desire to invest the brain with the elevated status of delivering compelling answers about human nature and intelligence" (2012, 118). The unproblematic nature of this ideal is reflected in Des Fitzgerald's study of autism research. Fitzgerald suggests that the complexity attributed to the brain prevents neuroscientists from developing straightforward explanations of mental disorders yet simultaneously puts them into a position in which they cannot deny the biological evidence that suggests a difference between "normal" and "pathological" brains. They are in an "awkward place: unable to tack autism to an emerging project of 'neuro-reductive' certainty [. . .] neuroscientists are equally unable to simply ignore both the evidence and the desire for a neurogenetic basis to the disorder" (Fitzgerald 2014, 247). In one way or another, the complexity attributed to the human brain must be dealt with within neuropsychiatry to reach explanations of mental disorders. My analysis of how diagnostic labels normatively structure how brain scans are interpreted, as well as how the fMRI is treated as allowing to map cognitive functions onto brain areas, can be considered an instance of how this "awkward place" is managed in practice. A large part of the work in neuropsychiatric practice seems precisely to handle being in this awkward place pragmatically.

One of the questions arising—precisely because of how brain imaging technologies are appropriated—when neuropsychiatric experiments become entangled with the behavior and interests of specific individuals is the extent to which such personalized "brained" experiments can be extrapolated beyond the medical biographies of specific individuals. In other words, a potential conflict seems to exist between the attempt to tailor scientific experiments to the mental disorders that individuals are diagnosed with and the desire to produce generalizable experimental results. This is a problem that Sally (the researcher discussing her experimental approach in excerpts 4–6) claims to have often encountered when presenting her research: aligning experiments to personal interests has often been treated as making the experiments "highly subjective" and hence "difficult to interpret." This indicates that the practice in which neuropsychiatric experiments are "brained" in the laboratory under study is not necessarily representative of the research domain of neuropsychiatry at large. However, what the present study points to is that the promise of objectivity associated with the introduction of brain imaging technologies in psychiatry constitutes an uneasy relationship between the complexity of the brain and the desired simplicity of psychiatric explanatory models. Rather than choosing between either a neuroreductionism or the means of traditional psychiatry, how brain imaging technologies are appropriated reveals a new dimension: (neuro-)biological evidence of mental disorders gives rise to a specific form of experimenting that highlights the inevitable subjectivity of mental disorders. In this sense, the (supposed) objectivity of scientific knowledge can be considered constitutive of how the human subject who is diagnosed with a certain mental disorder is understood.

In her discussion of how complexity is dealt with in synthetic biology, Talia Dan-Cohen distinguishes between *neglect* and *(purposive) ignorance*. She argues that in cases of "neglect," the complexity of biological systems is entirely removed from view, while ignorance "requires a carefully blinkered view and the ability to bide one's time, something that engineers, many of whom tethered their funding to technological promises, didn't build into their research programs" (Dan-Cohen 2016, 918). When applied to neuropsychiatry, the following can be said: in cases of neglect, it seems that neuropsychiatrists lose what has been gained through developments in brain science. In the case of ignorance, the complexity of the human brain becomes present as manageable and as something of which researchers acknowledge cannot be experimentally present as a whole.

The integration of fMRI in neuropsychiatry does make it less likely that the complexity of the brain is neglected, because complexity becomes perceptually available. However, this does not imply that a networked understanding of the human brain is straightforwardly applied to the explanation of mental disorders. Rather, fMRI can be understood as changing the hermeneutic process through which mental disorders are interpreted by adding a biological layer of interpretation that cannot be neglected. This hermeneutics is visible, for example, in how the development of experimental stimuli takes a different shape when mediated through fMRI. This can be considered a critical appropriation of complexity, such that it can be made to fit in—and made relative to—the aims of neuropsychiatric research and the promise neuropsychiatry bears for improving *individual* treatments in clinical practices. Against this background, how neuropsychiatric experiments are developed in this laboratory appears not as a case of straightforward neglect but rather as a form of ignorance that allows for managing the trade-off between the desired simplicity of neuropsychiatric explanations and the understanding of the brain as a complex organ stemming from the biological and neurological sciences.

In this chapter, I have conceptualized the constitution of the objects of neuropsychiatry in terms of how brain imaging technologies mediate the trade-off between complexity and simplicity. The epistemic concerns considered in the appropriation of technological mediation in neuropsychiatry are thus very different from those discussed in the previous chapter, which focused on the use of NIBS and NIBS-EEG in cognitive neuroscience. This suggests that how scientists appropriate technological mediations is contingent upon both the technologies used (e.g., NIBS vs. fMRI) and the pragmatic concerns at stake (e.g., cognitive neuroscience vs. neuropsychiatry). Accordingly, it does not make much sense at this point to say that brain imaging and brain stimulation technologies mediate the reality of "the brain," thereby giving rise to a specific way of understanding human behavior. Rather, brain imaging and brain stimulation technologies figure within broader processes of object constitution particular to the concerns of scientists who appropriate technological mediations in a specific practical setting. The novel ways in which these technologies disclose human behavior should, therefore, be understood in terms of how they allow the objectification of (aspects of) human behavior and cognition such that they become candidates for scientific investigation. This points to the postphenomenological approach of technological mediation being inasmuch a philosophy of *scientific practice* as a philosophy of *scientific instruments*.

NOTES

1. The term "neuropsychiatrist" should not be understood as referring to a clinical psychiatrist. None of the researchers at the laboratory under study were clinical psychiatrists themselves, but the participants in their experiments were recruited from the clinical psychiatrists at the same academic hospital.

2. The names of the participants have been altered, and the date and time indications have been deleted to minimize the possibility of identification.

3. Excerpts 1, 2, and 3 are parts of a conversation that originally occurred in Dutch. The excerpts are translated by the author. Original fragments can be delivered on request.

Conclusion

A Philosophy of Technological Mediations as a Philosophy of Scientific Practice

A guiding concern behind my exploration of technological mediations in scientific practice is clarifying the concept of "scientific practice." The adjective "scientific" suggests that there is something specific and distinctive about "scientific practices," while the noun "practice" indicates that there is, in one way or another, a continuity between scientific practices and other practices. How must the relationship between "scientific" and "practice" be understood such that a "scientific practice" is constituted? To address this problematic, the chapters in the first part of this book were also concerned with the extent to which a phenomenological analysis of the role of technologies in everyday life can be extended to the practices of science.

This book has further been guided by the following question: How does a new world—revealed through scientific instruments—enter the domain of human understanding, and what kind of world is this? As stated in the introduction, throughout this book, I have adhered to two interrelated assumptions: (i) an analysis of technologies (i.e., scientific instruments) should be a central aspect of a philosophy of science and (ii) a philosophy of science should focus on the practice of science rather than on scientific theorizing. In this conclusion, I show how the concepts of "mediating technologies" and "scientific practice" have been shaped in such a way that the questions posed in the introduction can be answered. Furthermore, I show the specific way in which the technologies used in neuroscientific practice shape the reality of the objects of the neurosciences.

THE MEDIATING ROLE OF TECHNOLOGIES IN SCIENTIFIC PRACTICE

Postphenomenological studies of scientific practice have been developed specifically in opposition to a *theory*-based understanding of science; instead, the explicit aim of postphenomenology is to develop a *technology*-driven account of science grounded in the practical interactions between scientists and technologies has been proposed (e.g., Ihde 1991, 100). Postphenomenologists often depart from the Heideggerian idea that our primary relation with the world is always a practical one with the tools and technologies around us, rather than one in which the entities around us appear as objects that are foreign to us. By centralizing the role of technologies in scientific practice, postphenomenology importantly counterbalances overly theoretically oriented philosophies of scientific instruments. When closely studying how scientists interact with technologies, it becomes clear that scientific instruments do not embody a univocal perspective (e.g., Giere 2006) that scientists can be acquainted with when having the correct training (e.g., Hacking 1983). Nor can it be said that scientific instruments embody specific knowledge claims that are waiting to be "found" by the scientists who use them (e.g., Baird 2004). Precisely what is at stake in a philosophy of scientific instruments—understood as mediating technologies—is the development of a collective epistemic stance toward what they allow to be observed and how these observations become meaningful in relation to the theoretical and practical concerns of scientific practices.

Postphenomenology, however, as I have argued in this book, sufficiently recognizes neither the importance of the theoretical aspects of science nor the collectivity of science. One aim of this book was to show not only the relevance of the philosophy of technology for the philosophy of science but also the shortcomings arising when a philosophy of technological mediation enters the domain of science. I have done so by reevaluating the insights of "classical philosophers," which posited a divide between philosophy and scientific research. Specifically, I have showed the ongoing relevance of the philosophy of Martin Heidegger and Gaston Bachelard by integrating their theoretical work into postphenomenology. I argued that postphenomenology has thus far too easily dismissed the Heideggerian insight that science explicitly intends to develop a relation with the world in terms of the present-at-hand. Accordingly, I have identified as a central concern of scientific investigations precisely how this relation is to be established, and suggested that a key task for a philosophy of scientific practice is analyzing how this is done.

With the introduction of the epistemology of Gaston Bachelard and his notion of phenomenotechnique, the question of whether a philosophy of science should focus on either theories or practices is replaced with the question of how scientific instruments co-constitute the concepts that scientists employ when understanding the reality that they study. This insight is relevant for a philosophy of technological mediation: technologies not only mediate the practical interactions with the things around us but also shape the terms that scientists use when attributing meaning to the reality that they investigate. I have argued that this indicates the necessity of studying the hermeneutic work arising within relations between scientists and scientific instruments, hermeneutic work that cannot be captured within Latour's work, which approaches scientific instruments semiotically and interprets them as inscription devices. This shift from a semiotic to a hermeneutic approach to the study of scientific practices is thus a consequence of what were identified as crucial insights of the philosophies of Heidegger and Bachelard.

Through this philosophical exploration, the scientific practices in which technologies play a mediating role are not solely understood in terms of the practical relations between scientists and technologies but are also extended in terms of the concepts and norms to which scientists orient (i.e., the empirical hermeneutics at work) when appropriating technological mediations. As such, an empirical philosophy of technoscience is necessary that is capable of revealing not only how scientists *act* in relation to technologies but also how technologies mediate how scientists *think*. When the latter is considered, it becomes possible to reveal which mode of thinking constitutes the specificity of a practice that can be characterized as "scientific." It is in this sense that a philosophy of technological mediation develops a specific concept of "scientific practice." Yet, how can one make researchable the ways in which technologies mediate the reality that scientists study on both a practical-perceptual and a theoretical-conceptual level?

To answer this question, it is necessary to not treat human-technology relations in isolation but to instead explicitly situate them in the practices within which they are constituted. Let me illustrate this by briefly reflecting on the case study on neuropsychiatry discussed in chapter 8: when beginning with the idea that scientific instruments mediate how and which scientific objects come into being, one is tempted to think that neuroscientists working with fMRI and NIBS herald such technologies for their revolutionary potential. Furthermore, when considering the public enthusiasm and optimism about the (cognitive) neurosciences as a source to describe the workings of the mind with scientific certainty by reducing it to the brain, one expects that this idea has its roots in the practical ways in which scientists engage with brain scans. When clinging to this perspective, it is equally tempting to expect that, if the optimism about the possibility to objectify the human mind is not explicitly present, as an analyst, one should attempt to unmask cognitive neuroscientists as secret reductionists.[1] In one way or another, fMRI and NIBS should do something new, must give rise to the unexpected, must change the worldview of the researchers. The microscope revolutionized biology; the telescope revolutionized astronomy; and now, fMRI and NIBS revolutionize the study of the human being. Beginning with this idea, it might come as a surprise that the actual practical routine of cognitive neuroscience does not match such expectations.

For example, when entering the neuropsychiatry laboratory in the case study discussed in chapter 8 for the first time, the principal investigator told me that the ultimate aim of the research conducted in the laboratory was to "show the inadequacy of the diagnostic categories in the DSM, and to develop a new model for psychiatry." Of course, this might have only been a casual remark, quickly uttered in between two more important meetings, but yet this was the type of revolutionary discourse one expects and deems important when attributing great significance to technological mediations. Surprisingly, from this perspective, the DSM and the categories within it neither did seem to play any significant role in the practical routine of the laboratory nor was the possibility of visualizing brain activity in vivo considered as a fundamental source of novel insights. Of course, the practical routine was simply there, and was—at least not visible to the novice observer—not ordered in relation to brain scans. The reality built in neuropsychiatric practice, accordingly, did not entirely coincide with the reality an analyst departing from the perspective of technological mediation would expect to find.

When specifically examining diagnostic categories and brain scans, technological mediations seem difficult to detect. This suggests that—if existing—these mediations often are not explicitly foregrounded but lurk in the background. The relevant questions then become which aspects of the background are to be foregrounded (i.e., cede from being background), and for what purposes is this done? Posing these questions already shows that from the start, the reality that is built within neuroscientific collectives cannot be cut loose from the broader practical aims and theoretical presuppositions (i.e., the phenomenotechnology) of the scientific practice under study.

A PHILOSOPHY OF SCIENTIFIC INSTRUMENTS AS A PHILOSOPHY OF SCIENTIFIC PRACTICE: TECHNOLOGICAL MEDIATION AS REALITY BUILDING

In the first chapter of this book, I argued that scientific practices are collective activities through and through and that understanding how technological mediations are appropriated by scientific collectives requires understanding how a shared epistemic stance with regard to specific technological mediations comes into being. This is why I suggest that *if* studying empirically the hermeneutics at work through which the reality that scientists study is constituted, (elements of) EM and CA should be integrated into a philosophy of technological mediation. The hermeneutic work to be studied—the practice under investigation—is, then, a study of the activities through which a shared reality is built in scientific practice and the meaning that is attributed to the objects appearing in it. The activity of reality building cannot be cut loose from human-technology relations but can be neither exhausted by them nor reduced to them. This endeavor marks the second augmentation of postphenomenology, which morphs a philosophy of technological mediation from a philosophy of scientific instruments into one of scientific practice.

How is the activity of reality building in scientific practice reflected in the empirical studies discussed in this book, and how does it enlarge our understanding of technological mediations? Applying philosophical concepts to concrete case studies helps to better understand not only the domain under investigation but also the philosophical concepts applied. In his essay *The Stranger*, Alfred Schütz describes how a newcomer becomes acquainted with the customs of a group that (s)he desires to become part of. The process of learning can only continue when the concepts and explanations of the newcomer do not coincide with those of the group. If the integration were complete, nothing would remain that could be explained or reflected upon. In Schütz's words, "if [the] process of inquiry succeeds, then [the group's] pattern and its elements will become to the newcomer a matter of course, an unquestionable way of life, a shelter, and a protection. But then the stranger is no stranger anymore, and his specific problems have been solved" (1944, 507). My suggestion is that something similar holds for the relation between the philosopher and the practices (s)he investigates: a divide always exists between the reality of the practice itself and the one ascribed to it by the analyzing philosopher.

Also, from a (post)phenomenological point of view, the position of a value-free, disinterested observer, who gives a neutral description of the practice (s) he engages with, cannot be assumed. In one way or another, the intentional relation with the practice under study is both already constituted by the prior interest in this practice and changes when the analyst immerses in it. Both the studying subject and the studied object are mutually shaped during the process of investigation. Hence, investigating a practice implies engaging with it, and the level of engagement shapes how the practice under study is described. As such, Schütz's words can serve as a cautionary remark that imposes a limit on the engagement with scientific practices: to some extent, one must remain strange to them to be able to lay bare the unquestioned schemes of reference of their members. Studying specific processes of reality building, after all, does not coincide with the building of this specific reality.

In EM, the purpose of the foregrounding of the background is directed at "the reform of technical reason" (Garfinkel 2002, 93). According to Garfinkel, the social sciences movement falsely assumes that no order exists in the activities of everyday life and that it is the task of sociology to create this order through developing formal rational procedures. He derives from this that the specific technical reasoning of the social and natural sciences is necessary to establish order into an otherwise disassociated and disordered set of practices. Ethnomethodological studies, however, "specif[y] the work of the social sciences and the natural sciences as naturally accountable sciences of practical action and practical reason" (Ibid. 93). This indicates that an order already exists prior to the intervention of technical reason. However, this preexisting order is not independent of technical reason: the specific way in which practices are ordered "includes the incarnate work by formal analysts [scientists]" (Ibid. 96). Thus, when foregrounding the practical actions in social science, the purpose of EM is to reveal a form of practical reason that precedes technical reason. The re-specification of the social and natural sciences, then, is an attempt to reveal the order already implied and accounted for in scientific practices and to make this order available for critical reflection.

Would pointing to scientific practices being structured around the appropriation of technological mediations, similar to the purpose of EM, be to point to something prior to the technical vocabulary of the science? Recent criticism of postphenomenological analyses suggests that this is not the case and argues that the predominantly descriptive character tacitly instantiates an overly optimistic stance toward technoscientific developments (e.g., Rao et al. 2015; Feenberg 2015; Romele 2017). Focusing too strongly on conducting empirical studies of technoscientific practices is then considered to undermine the potential of philosophy to develop a genuine critical position (e.g., Rao et al. 2015; Smith 2018).

THE MEDIATED REALITY OF NEUROSCIENTIFIC COLLECTIVES AND THE CRITICAL POTENTIAL OF A PHILOSOPHY OF TECHNOLOGICAL MEDIATION

How then, can a philosophy of technological mediation conduct empirical investigations without losing its critical potential? One way to combine the two is to conceive of empirical investigations as allowing to identify how the mode of thinking present in science can be understood as being constitutive of its objects. Laying bare this mode of this thinking—as I attempt to do in this book—allows for developing a critical relation to scientific practices and the objects constituted within them. This can be exemplified when applied to the neurosciences: departing from an ethnomethodological perspective, the neurosciences can be regarded as executing a form of technical reason, for example, by arguing for a change in educational programs (e.g., Carew and Magsamen 2010), psychiatric treatments (e.g., Insel and Cuthbert 2015), legal arrangements (cf. Meynen 2014), or marketing (cf. Ariely and Berns 2010). This is done with reference to a neurological order that is said to be constitutive of how human beings experience and act upon the world. The technical vocabulary of the neurosciences is, according to optimistic neuroscientists, necessary to show how the world is *actually* ordered, regardless of existing sociological and personal accounts of it (e.g., Lamme 2006; Soon et al. 2008).

Revealing something that is prior to technical or scientific reason can be a purpose of conducting laboratory studies. For example, Morana Alač's study on how neuroscientists interpret brain scans shows that interpretations are enacted through multimodal engagement within a specific social setting. She uses this finding to critically relate to how fMRI is generally perceived:

> Whereas fMRI technology presupposes that our cognitive processes can be reduced to the workings of an individual's brain, I argue for a view of cognition that is distributed between people and rooted not only in mental processes and computational inferences but also in a culturally shaped and socially enacted world. (Alač 2011, 20)

In this case, the empirical study of neuroscientific practice functions to counter a specific neuroscientific paradigm that accompanies the introduction of fMRI. Showing how the interpretation of brain scans presupposes the existence of a form of cognition that does not fit into the existing paradigm is, in Alač's work, a way to reveal the limits of a specific scientific paradigm. Revealing the *human* work involved in neuroscientific practice can—in line with Garfinkel's description of EM—be considered a re-specification of the neurosciences that allows to reveal an already existing form of reason.

What, then, is specific to focusing on the role of *technologies* in scientific practices and singling them out as doing something specific that must be acknowledged? Don Ihde has, as mentioned several times throughout this book, argued that any study of scientific practice *should* focus on material technologies, because these structure the interpretative processes at work. In his view, studying scientific practice therefore requires developing a material hermeneutics that focuses on the work of concrete technologies: besides human work and human reason, room should be made for the work of nonhumans and how these help disclosing reality. This shift of focus does not eliminate the importance of human work in scientific practice. However, it reconceptualizes the human work explicitly involved in terms of the appropriation of technological mediations (Verbeek 2016). As stated, I was inspired by this work to begin examining the hermeneutic processes that technologies instantiate in scientific practice by attempting to identify technological mediations and how these are appropriated.

In the two empirical studies discussed in this book, I have framed the appropriation of technological mediations in terms of the epistemic norms that neuroscientists orient themselves to when investigating the brain. Considering these norms suggests that scientist relate to scientific objects through not only technologies but also collective epistemic judgments. Although these norms can be conceptualized as appropriations of technological mediations, they only become visible in conversations between scientists in which experimental data is interpreted. Placing the appropriation of technological mediations at the center of scientific practice effectively counters the idea that technologies *determine* how scientific objects are constituted through a causal interaction of natural and material elements. On the contrary, by pointing to the contingencies involved during the appropriation of technological mediations, the human work involved in the constitution of scientific objects is rehabilitated, yet in a way that does not make it independent of the phenomenotechnological circumstances within which actions occur.

What follows from a focus on the human appropriation of technological mediations is that the investigations conducted in this book do not, strictly speaking, belong to a *material* hermeneutics (cf. Ihde 2009; Verbeek 2003)—at least not if "material" is assumed to refer to the material presence of concrete technologies. Rather, the "material" should be understood in terms of what I call a phenomenotechnology, consisting of both material technologies and immaterial norms and concepts. Foregrounding the epistemic norms that structure the reality of scientific objects in the cognitive neurosciences thus has consequences on how technological mediations are to be understood. Therefore, postphenomenological research should not only focus on how the physical characteristics of technologies shape how phenomena are perceived but also on how appropriations of technological mediations structure how the realities of these observed phenomena are conceived.

Let me clarify this with an example from the case study focusing on the constitution of visual attention. I argued that NIBS technologies—or, brain stimulation—mediate the reality of visual attention because they embody a strong causal norm toward which scientists orient themselves. Furthermore, I showed that the constitution of visual attention as an object that can be related to is the outcome of a trade-off between the norms of "causality" and "reality." The specific manner in which this trade-off was managed revealed that only when researchers can interpret their observations in terms of a recognizable causality does visual attention become real.[2] That scientists orient themselves to these norms indicates that their intentionality is hermeneutically structured not just by the material technologies but also by the normative expectations and goals of their investigations. To understand how technologies mediate the reality that scientists study, it is therefore necessary to focus on how these normative expectations manifest in relation to technologies and how the use of technologies takes shape in relation to the normative expectations. The constitution of visual attention is, accordingly, accomplished within a dialogue between material and immaterial hermeneutics. Through this dialogue, an intentional relation between scientists and visual attention comes into being—and it is within *this* intentional relation that NIBS becomes meaningful *as* causal.

Foregrounding this dialogue is not so much a re-specification of neuroscientific practice as it is a specification of the technical reasoning that actually occurs in scientific practice. A philosophy of technological mediation—by foregrounding the dialogue between material and immaterial hermeneutic processes—is an explication of the *technicity* of scientific practice. Because of its focus on technological mediations and their appropriations, a philosophy of technological mediation can be considered a *phenomenology of technicity*. Conceived in this manner, a conceptualization of the role of technologies in scientific practices reaches beyond the domain of embodied perception and shows how the constitution of scientific objects occurs within a specific mode of thought. Accordingly, laying bare the appropriation of technological mediations is done not by pointing to something *prior*—be it practical reasoning or multimodal engagement—to the technical reasoning of the (cognitive) neurosciences but rather to how scientists engage in technical thinking. Doing so allows to give both empirical and philosophical body to the claim that science is to be understood as *technoscience* (Ihde 2009; Latour 1987; Nordmann 2012).

This book has attempted to solicit an intuition in others regarding what it means to be alive with the neurosciences. I have done this by illuminating the processes within which these objects come into being within scientific laboratories and identifying which structuring roles epistemic norms have in this process. Intuitively grasping the normative contingencies involved in the constitution of neuroscientific objects helps neuroscientists reflect on their place in the technical ensemble that the laboratory comprises. Furthermore, this opens up the possibility to reflect on the relation between the technicification of human psychology and the specific conception of the human subject that accompanies it. An explication of the constitution of neuroscientific objects is, from this perspective, an explication of how it is to live with the cognitive neurosciences.

NOTES

1. Engaging in this project of unmasking might be triggered by recent developments in philosophy that aim to naturalize several branches of philosophy with reference to developments in neuroscientific research. For a prominent attempt to do so in the case of ethics, see, for example, Churchland (1986, 2018). For a criticism of such endeavors, see, for example, Choudhury and Slaby (2011), Vrecko (2010).

2. Now, of course, the fact that "reality" and "causality" normatively structure scientific investigations is unsurprising: the expectation of science is that it reveals causal relations between real objects. What is surprising is how these norms attain a specific meaning as appropriations of technological mediations.

Appendix

Transcription Symbols According to the Jefferson (2004) Transcription System

[ ]	indicate the start and end of overlapping speech
underlining	indicates emphasis
CAPITAL	indicate a section of speech noticeably louder than that surrounding it
he he	indicates voiced laughter
.hh	indicates in-breath
(1.1)	indicates pauses in seconds
(.)	indicates micro-pause, too short to measure
:	colons indicate (degree of) elongation of the prior sound
?	indicates question

Bibliography

Aagaard, Jesper. 2015. "Media Multitasking, Attention, and Distraction: A Critical Discussion." *Phenomenology and the Cognitive Sciences* 14 (4): 885–96.

Aagaard, Jesper. 2018. "An Introduction to Postphenomenological Methodologies". In *Postphenomenological Methodologies: New Ways in Mediating Techno-Human Relationships*, edited by Jesper Aagaard, Jan Kyrre Berg Friis, Jessica Sorenson, Oliver Tafdrup, and Cathrine Hasse, xi–xxx. London: Lexington Books.

Achterhuis, Hans (ed.). 2001. *American Philosophy of Technology: The Empirical Turn*. Translated by Robert P. Crease. Bloomington: Indiana University Press.

Akrich, Madeleine. 1992. "The De-Scription of Technical Objects." In *Shaping Technology/Building Society: Studies in Sociotechnical Change*, edited by Wiebe Bijker and John Law, 205–24. Cambridge: MIT Press.

Alač, Morana. 2008. "Working with Brain Scans: Digital images and Gestural Interaction in fMRI Laboratory." *Social Studies of Science* 38 (4): 483–508.

Alač, Morana. 2011. *Handling Digital Brains: A Laboratory Study of Multimodal Semiotic Interaction in the Age of Computers*. Cambridge: MIT Press.

Albus, James S., George A. Bekey, John H. Holland, Nancy G. Kanwisher, Jeffrey L. Krichmar, Mortimer Mishkin, Dharmendra S. Modha, et al. 2007. "A Proposal for a Decade of the Mind Initiative." *Science* 317 (5843): 1321.

Ali, Sabrina S., Michael Lifshitz, and Amir Raz. 2014. "Empirical Neuroenchantment: From Reading Minds to Thinking Critically." *Frontier in Human Neuroscience* 8 (357): 1–4.

Amso, Dima, and Gaia Scerif. 2015. "The Attentive Brain: Insights from Developmental Cognitive Neuroscience." *Nature Reviews Neuroscience* 16 (10): 606–19.

Ankeny, Rachel, Hasok Chang, Marcel Boumans, and Mieke Boon. 2011. "Introduction: Philosophy of Science in Practice." *European Journal for Philosophy of Science* 1: 303–7.

APA. 1994. *Diagnostic and Statistical Manual of Mental Disorders, 4th Edition: DSM-IV*. Washington, DC: American Psychiatric Association.

APA. 2013. *Diagnostic and Statistical Manual of Mental Disorders, 5th Edition: DSM-5*. Arlington: American Psychiatric Association.

Ariely, Dan, and Gregory S. Berns. 2010. "Neuromarketing: The Hope and Hype of Neuroimaging in Business." *Nature Reviews Neuroscience* 11: 284–92.

Bachelard, Gaston. 1949a. *Le Rationalisme Applique*. Paris: PUF.

Bachelard, Gaston. 1949b. "The Philosophical Dialectic of the Concepts of Relativity." Translated by Forrest W. Williams. In *Albert Einstein: Philosopher Scientist*, edited by Paul Arthur Schilpp, 563–80. New York: MJF Books.

Bachelard, Gaston. 1951. *L'Activité Rationaliste de la Physique Contemporaine*. Paris: PUF.

Bachelard, Gaston. 1964. *The Psychoanalysis of Fire*. Translated by Alan C. M. Ross. London: Routledge.

Bachelard, Gaston. 1968. *The Philosophy of No: A Philosophy of the New Scientific Mind*. Translated by G. C. Waterson. New York: Orion Press.

Bachelard, Gaston. 1984. *The New Scientific Spirit*. Translated by Arthur Goldhammer. Boston: Beacon Press.

Bachelard, Gaston. 1997. *Fragments of a Poetics of Fire*. Translated by Kenneth Haltman. Dallas: The Pegasus Foundation.

Bachelard, Gaston. 1999. *Water and Dreams: An Essay on the Imagination of Matter*. Translated by Edith R. Farrell. Dallas: The Pegasus Foundation.

Bachelard, Gaston. 2002. *The Formation of the Scientific Mind: A Contribution to a Psychoanalysis of Objective Knowledge*. Translated by Mary McAllister-Jones. Manchester: Clinamen Press.

Bachelard, Gaston. 2005a. "From *La Essai Sur la Connaissance Approchée*." Translated by Mary Tiles. In *Continental Philosophy of Science*, edited by Gary Gutting, 176–83. Oxford: Blackwell.

Bachelard, Gaston. 2005b. "Noumena and Microphysics." Translated by David Reggio. *Angelaki* 10 (2): 73–78.

Bachelard, Gaston. 2012. *The Flame of a Candle*. Translated by Joni Caldwell. Dallas: The Pegasus Foundation.

Baird, David. 2003. "Thing Knowledge: Outline of a Materialist Theory of Knowledge." In *The Philosophy of Scientific Experimentation*, edited by Hans Radder, 39–67. Pittsburgh: The University of Pittsburgh Press.

Baird, David. 2004. *Thing Knowledge: A Philosophy of Scientific Instruments*. Berkeley: University of California Press.

Beaulieu, Anne. 2002. "Images Are Not the (Only) Truth: Brain Mapping, Visual Knowledge, and Iconoclasm." *Science, Technology, & Human Values* 27 (1): 53–86.

Bloor, David. 1976. *Knowledge and Social Imagery*. London: Routledge.

Borck, Cornelius. 2012. "Toys Are Us: Models and Metaphors in Brain Research." In *Critical Neuroscience: A Handbook of the Social and Cultural Contexts of Neuroscience*, edited by Suparna Choudhury and Jan Slaby, 113–33. Oxford: Wiley-Blackwell.

Borck, Cornelius. 2016. "How We May Think: Imaging and Writing Technologies Across the History of the Neurosciences." *Studies in History and Philosophy of*

Science Part C: Studies in History and Philosophy of Biological and Biomedical Sciences 57: 112–20.

Brosnan, Caragh, and Mike Michael. 2014. "Enacting the 'Neuro' in Practice: Translational Research, Adhesion and the Promise of Porosity." *Social Studies of Science* 44 (5): 680–700.

Buckner, Randy L., Fenna M. Krienen, and B. T. Thomas Yeo. 2013. "Opportunities and Limitations of Intrinsic Functional Connectivity MRI." *Nature Neuroscience* 16 (7): 832–37.

Canguilhem, Georges. 1994. *Etudes d'Histoire et de Philosophie des Sciences Concernant les Vivant et la Vie*. Paris: Vrin.

Caputo, John. D. 2012. "Heidegger's Philosophy of Science: The Two Essences of Science." In *Heidegger on Science*, edited by Trish Glazebrook, 261–79. New York: SUNY Press.

Carew, Thomas J., and Susan H. Magsamen. 2010. "Neuroscience and Education: An Ideal Partnership for Producing Evidence-Based Solutions to Guide 21st Century Learning." *Neuron* 67 (5): 685–88.

Carusi, Annamaria, and Aud Sissel Hoel. 2014. "Towards a New Ontology of Vision." In *Representation in Scientific Practice Revisited*, edited by Catelijne Coopmans, Janet Vertesi, Michael E. Lynch, and Steve Woolgar, 201–22. London: The MIT Press.

Chang, Hasok. 2014. "Epistemic Activities and Systems of Practice: Units of Analysis in Philosophy of Science after the Practice Turn." In *Science After the Practice Turn in the Philosophy, History and Social Studies of Science*, edited by Léna Soler, Sjoerd Zwart, Michael Lynch, and Vincent Israel-Jost, 67–79. New York: Routledge.

Chimisso, Cristina. 2001. *Gaston Bachelard: Critic of Science and Imagination*. London: Routledge.

Chimisso, Cristina. 2008. "From Phenomenology to Phenomenotechnique: The Role of Early Twentieth-Century Physics in Gaston Bachelard's Philosophy." *Studies in History and Philosophy of Science Part A* 39 (3): 384–92.

Chimisso, Cristina. 2013. "The Life Sciences and French Philosophy of Science: Georges Canguilhem on Norms." In *New Challenges to Philosophy of Science*, edited by Hanne Andersen, Dennis Dieks, Wenceslao J. Gonzalez, Thomas Uebel, and Gregory Wheeler, 399–410. Dordrecht: Springer.

Chimisso, Cristina. 2015. "Narrative and Epistemology: Georges Canguilhem's Concept of Scientific Ideology." *Studies in History and Philosophy of Science* 54: 64–73.

Choudhury, Suparna, and Jan Slaby. 2012. "Introduction: Critical Neuroscience—Between Lifeworld and Laboratory." In *Critical Neuroscience: A Handbook of the Social and Cultural Contexts of Neuroscience*, edited by Suparna Choudhury and Jan Slaby, 1–26. Oxford: Wiley-Blackwell.

Churchland, Patricia S. 1986. *Neurophilosophy: Towards a Unified Science of the Mind-Brain*. Cambridge: The MIT Press.

Churchland, Patricia S. 2018. *Braintrust: What Neuroscience Tells Us About Morality*. New York: Princeton University Press.

Cohn, Simon R. 2012. "Disrupting Images: Neuroscientific Representations in the Lives of Psychiatric Patients." In *Critical Neuroscience: A Handbook of the Social and Cultural Contexts of Neuroscience*, edited by Suparna Choudhury and Jan Slaby, 179–93. Oxford: Wiley-Blackwell.

Coopmans, Catelijne, Janet Vertesi, Michael E. Lynch, and Steve Woolgar (eds). 2014. *Representation in Scientific Practice Revisited*. London: The MIT Press.

Dan-Cohen, Talia. 2016. "Ignoring Complexity: Epistemic Wagers and Knowledge Practices Among Synthetic Biologists." *Science, Technology, & Human Values* 41 (5): 899–921.

Daston, Lorraine, and Peter Galison. 2010. *Objectivity*, 2nd edition. New York: ZONE Books.

De Vries, Gerard. 2016. *Bruno Latour*. Cambridge: Polity Press.

Descartes, René. 1993. *Meditations on First Philosophy*, 3rd edition. Translated by Donald A. Cress. Indianapolis: Hackett Publishing Company.

Dumit, Joseph. 2004. *Picturing Personhood: Brain Scans and Biomedical America*. Princeton: Princeton University Press.

Edwards, Derek, Malcom Ashmore, and Jonathan Potter. 1995. "Death and Furniture: The Rhetoric, Politics, and Theology of Bottom Line Arguments Against Relativism." *History of the Human Sciences* 8 (2): 25–49.

Ellul, Jacques. 1964. *The Technological Society*. Translated by John Wilkinson. New York: Knopf.

Feenberg, Andrew. 1999. *Questioning Technology*. London: Routledge.

Feenberg, Andrew. 2015. "Making the Gestalt Switch." In *Postphenomenological Investigations: Essays on Human-Technology Relations*, edited by Robert Rosenberger and Peter-Paul Verbeek, 229–37. New York: Lexington Books.

Fitzgerald, Des. 2014. "The Trouble with Brain Imaging: Hope, Uncertainty and Ambivalence in the Neuroscience of Autism." *BioSocieties* 9 (3): 241–61.

Forss, Anette. 2012. "Cells and the (Imaginary) Patient: The Multistable Practitioner-Technology-Cell Interface in the Cytology Laboratory." *Medicine, Health Care and Philosophy* 15 (3): 295–308.

Foucault, Michel. 1989. "Introduction." Translated by Carolyn R. Fawcett. In *The Normal and the Pathological*, edited by Georges Canguilhem, 7–24. New York: ZONE Books.

Franssen, Maarten, Pieter E. Vermaas, Peter Kroes, and Anthonie Meijers (eds). 2016. *Philosophy of Technology After the Empirical Turn*. Dordrecht: Springer.

Friis, Jan K. B. O. 2012. "Perception: Embodiment and Beyond." *Foundations of Science* 17 (4): 363–67.

Garfinkel, Harold. 1952. *The Perception of the Other: A Study in Social Order*. Unpublished Ph.D. dissertation, Harvard University.

Garfinkel, Harold. 1967. *Studies in Ethnomethodology*. New Jersey: Prentice-Hall.

Garfinkel, Harold. 2002. *Ethnomethodology's Program: Working Out Durkheim's Aphorism*. Edited by Anne W. Rawls. New York: Rowman & Littlefield Publishers.

Garfinkel, Harold, and Harvey Sacks. 1986. "On Formal Structures of Practical Actions." In *Ethnomethodological Studies of Work*, edited by Harold Garfinkel, 160–93. London: Routledge.

Garfinkel, Harold, Michael Lynch, and Eric Livingston. 1981. "The Work of a Discovering Science Construed with Materials from the Optically Discovered Pulsar." *Philosophy of Social Science* 11 (2): 131–58.

Gest, Howard. 2004. "The Discovery of Microorganisms by Robert Hooke and Antoni van Leeuwenhoek, Fellows of the Royal Society." *Notes and Records of the Royal Society* 58 (2): 187–201.

Giere, Ronald. 2002. "Discussion Note: Distributed Cognition in Epistemic Cultures." *Philosophy of Science* 69 (4): 637–44.

Giere, Ronald. 2006. *Scientific Perspectivism.* Chicago: University of Chicago Press.

Giere, Ronald, and Barton Moffatt. 2003. "Distributed Cognition: Where the Cognitive and the Social Merge." *Social Studies of Science* 33 (2): 301–10.

Gilbert, Nigel, and Michael Mulkay. 1984. *Opening Pandora's Box: A Sociological Analysis of Scientist's Discourse*. Cambridge: Cambridge University Press.

Glazebrook, Trish. 2000a. "From *φύσις* to Nature, *τέχνη* to Technology: Heidegger on Aristotle, Galileo, and Newton." *The Southern Journal of Philosophy* 38: 95–118.

Glazebrook, Trish. 2000b. *Heidegger's Philosophy of Science*. New York: Fordham University Press.

Greicius, Michael. 2008. "Resting-State Functional Connectivity in Neuropsychiatric Disorders." *Current Opinion in Neurology* 21 (4): 424–30.

Gross, Sky. 2009. "Experts and'Knowledge that Counts': A Study into the World of Brain Cancer Diagnosis." *Social Science & Medicine* 69 (12): 1819–26.

Hacking, Ian. 1983. *Representing and Intervening: Introductory Topics in the Philosophy of Natural Science*. Cambridge: Cambridge University Press.

Harman, Graham. 2004. *Prince of Networks: Bruno Latour and Metaphysics*. Melbourne: Re.press.

Hasse, Cathrinne. 2008. "Postphenomenology: Learning Cultural Perception in Science." *Human Studies* 31: 43–61.

Heiberg Engel, Peter Johan. 2008. "Tacit Knowledge and Visual Expertise in Medical Diagnostic Reasoning: Implications for Medical Education." *Medical Teacher* 30 (7): 184–88.

Heidegger, Martin. 1977a. "The Question Concerning Technology." Translated by William Lovitt. In *The Question Concerning Technology and Other Essays*, edited by William Lovitt, 3–35. London: Garland Publishing.

Heidegger, Martin. 1977b. "The Age of the World Picture." Translated by William Lovitt. In *The Question Concerning Technology and Other Essays*, edited by William Lovitt, 115–54. London: Garland Publishing.

Heidegger, Martin. 1977c. "Science and Reflection." Translated by William Lovitt. In *The Question Concerning Technology and Other Essays*, edited by William Lovitt, 155–82. London: Garland Publishing.

Heidegger, Martin. 1985. *Being and Time*. Translated by John Macquarrie and Edward Robinson. Oxford: Basil Blackwell.

Heidegger, Martin. 1992. *Parmenides*. Translated by André Schuwer and Richard Rojcewicz. Indianapolis: Indiana University Press.

Heritage, John. 1984. *Garfinkel and Ethnomethodology*. Cambridge: Polity Press.

Herrmann, Cristoph S., Daniel Strüber, Randolph F. Helfrich, and Andreas K. Engel. 2016. "EEG Oscillations: From Correlation to Causality." *International Journal of Psychophysiology* 103: 12–21.

Hooke, Robert. 1665. *Micrographia: Or Some Physiological Description of Minute Bodies Made by Magnifiying Glasses. With Observations and Inquiries Thereupon.* London: J. Martyn and J. Allestry.

Horkheimer, Max, and Theodor W. Adorno. 1969. *Dialektik der Aufklärung: Philosophische Fragmente*. Frankfurt: Fischer.

Høstaker, Roar. 2005. "Latour – Semiotics and Science Studies." *Science Studies* 18 (2): 2–25.

Hutchins, Edward. 1995. *Cognition in the Wild.* Cambridge: MIT Press.

Hyder, David. 2003. "Foucault, Cavaillès, and Husserl on the Historical Epistemology of the Sciences." *Perspectives on Science* 10 (4): 107–29.

Hyman, Steven E. 2010. "The Diagnosis of Mental Disorders: The Problem of Reification." *Annual Review of Clinical Psychology* 6: 155–79.

Ihde, Don. 1979. *Technics and Praxis*. Dordrecht: D. Reidel Publishing Company.

Ihde, Don. 1990. *Technology and the Lifeworld: From Garden to Earth.* Indianapolis: Indiana University Press.

Ihde, Don. 1991. *Instrumental Realism: The Interface Between Philosophy of Science and Philosophy of Technology*. Indianapolis: Indiana University Press.

Ihde, Don. 1993. *Postphenomenology: Essays in the Postmodern Context*. Evanston: Northwestern University Press.

Ihde, Don. 1997. "The Structure of Technology Knowledge." *International Journal of Technology Design Education* 7: 73–79.

Ihde, Don. 1998. *Expanding Hermeneutics: Visualism in Science*. Evanston: Northwestern University Press.

Ihde, Don. 2009. *Postphenomenology and Technoscience: The Peking Lectures*. New York: The State University of New York Press.

Ihde, Don. 2010. *Heidegger's Technologies: Postphenomenological Perspectives*. New York: Fordham University Press.

Ihde, Don, and Evan Selinger. 2004. "Merleau-Ponty and Epistemology Engines." *Human Studies* 27 (4): 361–76.

Insel, Thomas R., and Bruce N. Cuthbert. 2015. "Brain Disorders? Precisely." *Science* 348 (6234): 499–500.

Jaspers, Karl. 1951. *Man in the Modern Age*. Translated by Eden Paul and Cedar Paul. London: Routledge & Kegan Paul.

Kaplan, Edward K. 1972. "Gaston Bachelard's Philosophy of Imagination: An Introduction." *Philosophy and Phenomenological Research* 33 (1): 1–24.

Kletzl, Sebastian. 2014. "Scrutinizing Thing Knowledge." *Studies in History and Philosophy of Science Part A* 47: 118–23.

Knorr-Cetina, Karin. 1999. *Epistemic Cultures: How the Sciences Make Knowledge.* London: Harvard University Press.

Knorr-Cetina, Karin, and Michael Mulkay (eds). 1983. *Science Observed: Perspectives on the Social Study of Science*. London: SAGE.

Koch, Christof, Clay Reid, Hongkui Zeng, Stefan Mihalas, Mike Hawrylycz, John Philips, Chinh Dang, and Allan Jones. 2015. "Project Mindscope." In *The Future of the Brain: Essays by the World's Leading Neuroscientists*, edited by Gary Marcus and Jeremy Freeman, 25–39. Princeton: Princeton University Press.

Kochan, Jeff. 2010. "Latour's Heidegger." *Social Studies of Science* 40 (4): 579–98.

Koschmann, Timothy, and Alan Zemel. 2014. "Instructed Objects." In *Interacting with Objects: Language, Materiality, and Social Activity*, edited by Maurice Nevile, Pentti Haddington, Trine Heinemann, and Mirka Rauniomaa, 357–77. Amsterdam: John Benjamins.

Kroes, Peter A., and Anthonie Meijers (eds). 2000. *The Empirical Turn in the Philosophy of Technology*. Amsterdam: JAI.

Lamme, Victor. 2006. "Towards a True Neural Stance on Consciousness." *Trends in Cognitive Sciences* 10 (11): 494–501.

Latour, Bruno. 1983. "Give Me a Laboratory and I Will Raise the World." In *Science Observed*, edited by Karin Knorr-Cetina and Michael Mulkay, 141–70. London: SAGE.

Latour, Bruno. 1986. "Visualization and Cognition: Thinking with Eyes and Hands." *Knowledge and Society* 6: 1–40.

Latour, Bruno. 1987. *Science in Action: How to Follow Scientists and Engineers Through Society*. Cambridge: Harvard University Press.

Latour, Bruno. 1988a. "The Enlightenment Without the Critique: A Word on Michel Serres' Philosophy." In *Contemporary French Philosophy*, edited by A. Philips Griffiths, 83–98. Cambridge: Cambridge University Press.

Latour, Bruno. 1988b. *The Pasteurization of France*. Translated by Alan Sheridan and John Law. Cambridge: Harvard University Press.

Latour, Bruno. 1993a. "Pasteur on Lactic Acid Yeast: A Partial Semiotic Analysis." *Configurations* 1 (1): 129–46.

Latour, Bruno. 1993b. *We Have Never Been Modern*. Translated by Catherine Porter. Cambridge: Harvard University Press.

Latour, Bruno. 1994. "On Technical Mediation – Philosophy, Sociology, Genealogy." *Common Knowledge* 3 (2): 29–64.

Latour, Bruno. 1999a. *Pandora's Hope – Essays in the Reality of Science Studies*. Cambridge: Harvard University Press.

Latour, Bruno. 1999b. "On Recalling ANT." In *Actor Network and After*, edited by John Law and John Hassard, 15–25. Oxford: Blackwell.

Latour, Bruno. 2004. "Why Has Critique Run Out of Steam? From Matters of Fact to Matters of Concern." *Critical Inquiry* 30: 225–48.

Latour, Bruno. 2005. *Reassembling the Social: An Introduction to Actor-Network Theory*. Oxford: Oxford University Press.

Latour, Bruno. 2008. *What Is the Style of Matters of Concern?* Assen: Van Gorcum.

Latour, Bruno, and Steve Woolgar. 1986. *Laboratory Life: The Construction of Scientific Facts*, 2nd edition. Princeton: Princeton University Press.

Law, John. 2007. "Actor Network Theory and Material Semiotics." In *The New Blackwell Companion to Social Theory*, 3rd edition, edited by Bryan S. Turner, 141–58. Oxford: Blackwell.

Lichtenstein, Paul, Eva Carlström, Maria Rastam, Christopher Gillberg, and Hendrik Anckarsäter. 2010. "The Genetics of Autism Spectrum Disorders and Neuropsychiatric Disorders in Childhood." *American Journal of Psychiatry* 67: 1357–63.

Littlefield, Melissa. 2009. "Constructing the Organ of Deceit: The Rhetoric of fMRI and Brain Fingerprinting in Post-9/11 America." *Science, Technology, & Human Values* 34 (3): 365–92.

Littlefield, Melissa, and Jenell M. Johnson (eds). 2012. *The Neuroscientific Turn: Transdisciplinarity in the Age of the Brain*. Michigan: The University of Michigan Press.

Loscerbo, John. 1981. *Being and Technology: A Study in the Philosophy of Martin Heidegger*. The Hague: Martinus Nijhoff Publishers.

Lynch, Michael. 1985. *Art and Artifact in Laboratory Science: A Study of Shop Work and Shop Talk in a Research Laboratory*. London: Routledge & Kegan Paul.

Lynch, Michael. 1988. "The Externalized Retina: Selection and Mathematization in the Visual Documentation of Objects in the Life Sciences." *Human Studies* 11 (2): 201–34.

Lynch, Michael. 1993. *Scientific Practice and Ordinary Action: Ethnomethodology and Social Studies of Science*. Cambridge: Cambridge University Press.

Lynch, Michael. 1994. "Representation Is Overrated: Some Critical Remarks About the Use of the Concept of Representation in Science Studies." *Configurations* 2 (1): 137–49.

Lynch, Michael, and David Bogen. 1994. "Harvey Sacks's Primitive Natural Science." *Theory, Culture & Society* 11: 65–104.

Maung, Hane H. 2016. "Diagnosis and Causal Explanation in Psychiatry." *Studies in History and Philosophy of Biological and Biomedical Sciences* 60: 15–24.

McManus, Denis. 2012. *Heidegger & the Measure of Truth*. Oxford: Oxford University Press.

Menon, Vinod. 2011. "Large-Scale Brain Networks and Psychopathology: A Unifying Triple Network Model." *Trends in Cognitive Sciences* 15 (1): 483–506.

Merleau-Ponty, Maurice. 2012. *Phenomenology of Perception*. Translated by Donald A. Landes. London: Routledge.

Meynen, Gerben. 2014. "Neurolaw: Neuroscience, Ethics, and Law: Review Essay." *Ethical Theory and Moral Practice* 17 (4): 819–29.

Miniussi, Carlo, Justin A. Harris, and Manuela Ruzzoli. 2013. "Modelling Non-Invasive Brain Stimulation in Cognitive Neuroscience." *Neuroscience and Biobehavioral Reviews* 37 (8): 1702–12.

Nevile, Maurice, Pentti Haddington, Trine Heinemann, and Mirka Rauniomaa (eds). 2014. *Interacting with Objects: Language, Materiality, and Social Activity*. Amsterdam: John Benjamins.

Nordmann, Alfred. 2012. "Another Parting of the Ways: Intersubjectivity and the Objectivity of Science." *Studies in History and Philosophy of Science* 43: 38–46.

Ogawa, S., T. M. Lee, A. R. Kay, and D. W. Tank. 1990. "Brain Magnetic Resonance Imaging with Contrast Dependent Blood Oxygenation." *Proceedings of the National Academy of Sciences* 87 (24): 9868–72.

Olesen, Finn. 2012. "Scientific Objectivity and Postphenomenological Perception." *Foundations of Science* 17 (4): 357–62.

Pickersgill, Martyn. 2011. "'Promising' Therapies: Neuroscience, Clinical Practice, and the Treatment of Psychopathy." *Sociology of Health & Illness* 33 (3): 448–64.

Pickersgill, Martyn, and Ira Van Keulen (eds). 2012. *Sociological Reflections on the Neurosciences*. Bingley: Emerald.

Pitt, Joseph. 2007. "Speak to Me: Essay Review of *Thing Knowledge: A Philosophy of Scientific Instruments* by Davis Baird." *Metascience* 16 (1): 51–59.

Poldrack, Russel A., Jeanette A. Mumford, Tom Schonberg, Donald Kalar, Bishal Barman, and Tal Yarkoni. 2012. "Discovering Relations Between Mind, Brain, and Mental Disorders Using Topic Mapping." *PLOS Computational Biology* 8 (10): e1002707.

Popper, Karl. 1959. *The Logic of Scientific Discovery*. London: Hutchinson & Co.

Rao, Mithun B., Joost Jongerden, Pieter Lemmens, and Guido Ruivenkamp. 2015. "Technological Mediation and Power: Postphenomenology, Critical Theory, and Autonomist Marxism." *Philosophy and Technology* 28 (3): 449–74.

Rees, Tobias. 2016. *Plastic Reason: An Anthropology of Brain Science in Embryogenetic Terms*. Oakland: University of California Press.

Rheinberger, Hans-Jörg. 1997. "Experimental Complexity in Biology: Some Epistemological and Historical Remarks." *Philosophy of Science* 64 (4): S245–54.

Rheinberger, Hans-Jörg. 2005. "Gaston Bachelard and the Notion of "Phenomenotechnique."" *Perspectives on Science* 13 (3): 313–28.

Rheinberger, Hans-Jörg. 2010. *An Epistemology of the Concrete: Twentieth-Century Histories of Life*. Durham: Duke University Press.

Romele, Alberto. 2017. "Entangled in Digital Media." *Techné: Research in Philosophy and Technology* 21 (1): 106–13.

Rosenberger, Robert. 2008. "Perceiving Other Planets: Bodily Experience, Interpretation, and the Mars Orbiter Camera." *Human Studies* 31: 63–75.

Rosenberger, Robert. 2011. "A Case Study in Applied Philosophy of Imaging: The Synaptic Vesicle Debate." *Science, Technology, & Human Values* 36 (1): 6–32.

Rosenberger, Robert. 2013. "Mediating Mars: Perceptual Experience and Scientific Imaging Technologies." *Foundations of Science* 18: 75–91.

Rosenberger, Robert. 2017. "Notes on a Nonfoundational Phenomenology of Technology." *Foundations of Science* 22 (3): 471–94.

Rosenberger, Robert, and Peter-Paul Verbeek (eds). 2015. *Postphenomenological Investigations: Essays on Human-Technology Relations*. London: Lexington Books.

Roskies, Adina L. 2007. "Are Neuroimages like Photographs of the Brain?" *Philosophy of Science* 74 (5): 860–72.

Rouse, Joseph. 1996. *Engaging Science: How to Understand Its Practices Philosophically*. New York: Cornell University Press.

Sack, Alexander. 2006. "Transcranial Magnetic Stimulation, Causal-Structure-Function Mapping and Networks of Functional Relevance." *Current Opinion in Neurobiology* 16 (5): 593–99.

Sacks, Harvey. 1992. *Lectures on Conversation Volumes I & II*. Edited by Gail Jefferson. Oxford: Blackwell.

Schaffer, Simon. 1992. "Essay Review: The Eighteenth Brumaire of Bruno Latour." *Studies in History and Philosophy of Science* 22 (1): 174–92.

Schmidgen, Henning. 2014. *Bruno Latour in Pieces: An Intellectual Biography*. Translated by Gloria Custance. New York: Fordham University Press.

Schütz, Alfred. 1944. "The Stranger: An Essay in Social Psychology." *American Journal of Sociology* 49 (4): 499–507.

Shapin, Steven, and Simon Schaffer. 2011. *Leviathan and the Air-Pump: Hobbes, Boyle, and the Experimental Life*, 2nd edition. Princeton: Princeton University Press.

Smith, Dominic. 2018. *Exceptional Technologies: A Continental Philosophy of Technology*. London: Bloomsbury.

Soon, Chun S., Marcel Brass, Hans-Jochen Heinze, and John-Dylan Haynes. 2008. "Unconscious Determinants of Free Decisions in the Human Brain." *Nature Neuroscience* 11: 543–45.

Sporns, Olaf. 2014. "Contributions and Challenges for Network Models in Cognitive Neuroscience." *Nature Neuroscience* 17 (2): 652–60.

Suárez, Mauricio. 2004. "An Inferential Conception of Scientific Representation." *Philosophy of Science* 71 (5): 767–79.

Svenaeus, Fredrik. 2013. "The Relevance of Heidegger's Philosophy of Technology for Biomedical Ethics." *Theory of Medicine and Bioethics* 34: 1–15.

Te Molder, Hedwig. 2015. "What Happened to Post-Cognitive Psychology?" In *Discursive Psychology: Classic and Contemporary Issues. Explorations in Social Psychology*, edited by Cristian Tileaga and Elizabeth Stokie, 87–100. New York: Routledge.

Te Molder, Hedwig, and Jonathan Potter (eds). 2005. *Conversation and Cognition*. Cambridge: Cambridge University Press.

Ten Have, Paul. 1990. "Methodological Issues in Conversation Analysis." *Bulletin de Methodologie Sociologie* 27: 23–51.

Van Baalen, Sophie, Annamaria Carusi, Ian Sabroe, and David G. Kiely. 2017. "A Social-Technological Epistemology of Clinical Decision-Making as Mediated by Imaging." *Journal of Evaluation in Clinical Practice* 23 (5): 949–58.

Van Fraassen, Bas C. 1980. *The Scientific Image*. Oxford: Clarendon Press.

Van Fraassen, Bas C. 2008. *Scientific Representation: Paradox of Perspective*. Oxford: Clarendon Press.

Verbeek, Peter-Paul. 2005. *What Things Do*. Pennsylvania: Pennsylvania University Press.

Verbeek, Peter-Paul. 2011. *Moralizing Technology: Understanding and Designing the Morality of Things*. Chicago: The University of Chicago Press.

Verbeek, Peter-Paul. 2016. "Toward a Theory of Technological Mediation: A Program for Postphenomenological Research." In *Technoscience and Postphenomenology:*

The Manhattan Papers, edited by Jan K. B. Friis and Robert P. Crease, 189–204. London: Lexington Books.

Vertesi, Janet. 2012. "Seeing Like a Rover: Visualization, Embodiment, and Interaction on the Mars Exploration Rover Mission." *Social Studies of Science* 42 (3): 393–414.

Vidal, Fernando. 2009. "Brainhood, Anthropological Figure of Modernity." *History of the Human Sciences* 22 (1): 5–36.

Visser, Gerard. 2014. *Heideggers Vraag naar de Techniek*. Nijmegen: Uitgeverij Vantilt.

Vrecko, Scott. 2010. "Neuroscience, Power and Culture: An Introduction." *History of the Human Sciences* 23 (1): 1–10.

Walsh, Vincent, and Alan Cowey. 2000. "Transcranial Magnetic Stimulation and Cognitive Neuroscience." *Nature Reviews Neuroscience* 1 (1): 73–79.

Ward, Lawrence M. 2003. "Synchronous Neural Oscillations and Cognitive Processes." *Trends in Cognitive Science* 7 (12): 553–59.

Whitehead, Alfred N. 1920. *The Concept of Nature*. Cambridge: Cambridge University Press.

Wittenburg, Peter, Hennie Brugman, Albert Russel, Alex Klassmann, and Han Sloetjes. 2006. "ELAN: A Professional Framework for Multimodality Research." In *Proceedings of LREC 2006, Fifth International Conference on Language Resources and Evaluation*, Paris: ELRA. 1556–59.

Wolf, Thomas. 2003. "Konstitution und Kritik der Wissenschaften bei Heidegger." *Zeitschrift für Philosophische Forschung* 57 (1): 94–110.

Wolf, Thomas. 2005. *Hermeneutik und Technik: Martin Heideggers Auslegung des Lebens und der Wissenschaft als Answort auf die Krise der Moderne*. Würzburg: Königshausen & Neumann.

Zwier, Jochem, Vincent Blok, and Pieter Lemmens. 2016. "Phenomenology and the Empirical Turn: A Phenomenological Analysis of Postphenomenology." *Philosophy & Technology* 29: 313–33.

Index

About the Author

Bas de Boer is a philosopher of technoscience working as a postdoctoral researcher at the University of Twente, The Netherlands. He is interested in how technoscientific developments shape the ways in which people understand themselves and the world around them.

www.ingramcontent.com/pod-product-compliance
Ingram Content Group UK Ltd.
Pitfield, Milton Keynes, MK11 3LW, UK
UKHW041842031026
15587UKWH00002B/956

9 781793 627841